RECRYSTALLIZATION IN MATERIALS PROCESSING

Edited by **Vadim Glebovsky**

Recrystallization in Materials Processing
http://dx.doi.org/10.5772/58713
Edited by Vadim Glebovsky

Contributors

Renu Chadha, Kunal Chadha, Maninder Karan Vashisht, Yashika Bhalla, Vadim Glebovsky, Yousef Javadzadeh, Solmaz Maleki Dizaj, Zhila Vazifehasl, Masumeh Mokhtarpoor, Yuriy Perlovich, Margarita Isaenkova, Dong Nyung Lee, Sung Bo Lee, Sergei Mileiko

Published by InTech

Janeza Trdine 9, 51000 Rijeka, Croatia

Publishing Process Manager Iva Simcic
Technical Editor InTech DTP team
Cover InTech Design team

First published October, 2015
Printed in Croatia

Additional hard copies can be obtained from orders@intechopen.com

Recrystallization in Materials Processing, Edited by Vadim Glebovsky
p. cm.
ISBN 978-953-51-2196-1

For EU product safety concerns:
IN TECH d.o.o., Prolaz Marije Krucifikse Kozulić 3, 51000 Rijeka, Croatia,
info@intechopen.com or visit our website at intechopen.com.

Meet the editor

Vadim Glebovsky is a professor at the Institute of Solid State Physics, Russian Academy of Sciences in Chernogolovka, Moscow district, Russia. He graduated at the Baikov Institute for Metallurgy, Russian Academy of Sciences, Moscow, Russia. Then he worked successively at the Institute of Solid State Physics, Russian Academy of Sciences in Chernogolovka. For the long periods of time he stayed abroad for the scientific studies at the Max-Planck-Institutes fuer Metallforschung in Stuttgart and at the Technical University in Eindhoven. His researches span various areas of the physical metallurgy of high-purity refractory metals and alloys. In particular, he is interested in both theory and practice of growing single crystals and bicrystals of transition metals, studying the crystallographically related surface properties of high-purity refractory metals of 4, 5 and 6 groups of the Periodic Table. He is an expert in the electron-beam floating zone melting technique as well in the various high-temperature processing of single-crystalline and polycrystalline high-purity refractory metals and alloys. Professor Vadim Glebovsky is an author of over 300 research papers, 75 patents and several monographs on levitation melting, single crystals growing and other subjects of materials science.

Contents

Preface

Recrystallization is occurred in crystalline materials. However, it is mostly inherent for metals, so the bulk of the research of recrystallization is carried out on metals. There is a large collection of reviews, monographs and research papers are widely known on this subject for the period of hundred years of studies. During this time, a considerable progress has been done, both in the knowledge of recrystallization and relied phenomena. Driving torques of research in this area were primarily industrial applications, because they need not only new technologies but also the creation of metallic materials with optimal physical and mechanical parameters for new application areas. Thus, the high demands are appeared for development of physical models describing the metal-forming processes and optimizing the microstructure and macrostructure of the metallic products. Remarkable advances have been made in recent years in the science and technology of recrystallization. The further development of this science due to the theoretical and experimental studies is still among main tasks of this extraordinary phenomenon - recrystallization. If it would be taken into account that recrystallization began to seize new areas such as chemistry and pharmacology, gradually one come to a conclusion that the recrystallization science may be inexhaustible, eternal.

It was our aim to put together reviews and papers of recrystallization and relayed processes and then to evaluate them from one viewpoint. We made emphasis on the practical use of the processes of recrystallization to give technological guidelines for their application, a solid guide to the monographs and papers, and an overview of each process included in recrystallization. The authors understood that it is impossible to write a book on recrystallization, because it is an already more or less defined subject in material science and technology; however, the most part of aspects of it is not yet well understood and the experimental evidence is often poor and contradictory. It would have been desirable to quantify all aspects of the phenomena of recrystallization and to develop the theories from first principles. However, this is not yet possible, and the reader will find within this book a mixture of relatively sound theory, reasonable assumptions and speculations. Unfortunately, recovery and recrystallization are not still understood in depth because of poor understanding the nature of the deformed state. Recovery and recrystallization depend on the nature of the deformed state and involve the formation, removal and movement of grain boundaries.

This book is also dedicated to recrystallization and opens with chapter 1 that provides an interesting study of the behavior of a number of metallic materials (copper, silver, nickel alloys, etc.) deposited by electroless, in the form of wires and thin films. It shows the behavior of grains during annealing depending on the temperature and annealing. Two next chapters are connected with single crystals of refractory metals (tungsten, molybdenum, zirconium). In chapter 2 on tungsten and molybdenum, recrystallization is used as a technological trick to getting these metals with high technological plasticity. Since high plasticity is also a function of the impurity concentrations (mainly the content of interstitials), the author has developed in parallel the model of the pair interaction of carbon and oxygen and their removal of refractory metals. I believe that the added value of this chapter is precisely detailed analysis of the physical behavior of these impurities in the refractory metals. Based on this analysis, the author ranks the possible purification of refractory metals from interstitials, which is of the great interest to researchers. Chapter 3 presents a continuation of the theme, well-developed in the book "Recrystallization" published by InTech in 2011. This chapter examines the processes of recrystallization of the crystalline and polycrystalline alpha zirconium, which certainly makes our knowledge in this area. Experimental methods used in the study, almost the same, so that the reader, turning once attention to chapter of these authors, published in 2011, can easily continue reading the story of their studies. I would like to draw attention to the fact that, despite of the important role of zirconium in specific areas, there is still a great shortage of information regarding the optimal conditions of plastic deformation as well the optimal microstructures of rolled zirconium. The chapter written by Professor S. Mileiko presents a new class of materials – composites made of metal and oxide materials. The internal crystallisation technique to produce model composites as well as single crystalline and eutectic oxide fibres has been invented in the author's laboratory. Now this technique is considered as a base for developing effective technologies for production of a large family of oxide fibres with mechanical properties similar to those of the fibres produced by using traditional methods. ICM fibres can be used to obtain heat resistant composites with various matrices including nickel superalloys and high entropy alloys. An appropriate choice of the chemical composition of oxide reinforcements for refractory matrices can be a way to reach a necessary balance of the main properties of heat resistant materials, those being creep resistance, fracture toughness and oxidation resistance. Composite ICM fibres containing highly anisotropic oxide inclusions have been obtained and are considered as effective reinforcements for brittle matrices to produce composites with sufficiently high values of fracture toughness.

Next two chapters (5 and 6) are about specific crystallization and recrystallization features in drugs and pharmacology. Both chapters are of high interest to read about a new approach to such old world crafts as cooking food and preparing medicine. It can be refreshed in the memory of readers the great saying of Hippocrates "Let food be thy medicine and medicine be thy food". The chapters are really interesting and useful, even to simple readers. The authors prepared a competent and comprehensive overview of the main ideas and achievements in these fields of medicine and pharmacy.

It is supposed that writing a book of this kind is necessary to provide a modern description of recrystallization, and in which newly original or important up-to-date achievements have been done. However, a more experienced reader knows, that the value of such books is not only expanding our knowledge of the complex phenomena, but also give us a lot of advices for solving various technological problems.

I believe that the book "Recrystallization in Materials Processing" will be useful both to those who wish to gain a broad perspective on materials processes in general and to specialists using these processes. Finally, I wish to thank the contributing authors for their considerable efforts in summarizing the voluminous ideas and literature of their specialties. I would also thank our families and close friends for their great patience and benevolence, that they showed during our writing and preparation of this book.

Prof. Vadim Glebovsky
Institute of Solid State Physics
Russian Academy of Sciences, Russia

Chapter 1

Abnormal Grain Growth Texture

Dong Nyung Lee and Sung Bo Lee

Additional information is available at the end of the chapter

http://dx.doi.org/10.5772/59724

1. Introduction

When the dislocation density in fabricated materials is not high enough to cause recrystallization, a theory for the texture evolution during recrystallization (the strain-energy-release maximization theory) [1] is not applicable. In this case, grain boundaries, free surfaces, and internal stresses can be sources of internal energy and will control the evolution of textures during annealing. For bulk materials, grain boundary energy dominates internal energy. For polycrystalline films, grain boundaries, surface, interface, and internal stress can be sources of internal energy. Minimization of all the energies can, in principle, determine the competition between the variously oriented and stressed growing grains, which in turn gives rise to the abnormal grain growth texture [2]. In this chapter, the evolution of abnormal grain growth textures in drawn silver and copper wires, nanocrystalline deposits such as Fe-Ni alloy electrodeposits, Ni electrodeposits, electroless Ni-P deposits, electroless Ni-Co-P deposits, electroless Ni-Cu-P deposits, and copper interconnects will be discussed. Textures of Al-1%Cu interconnects and annealing textures of nanocrystalline copper thin films will also be discussed.

2. The evolution of abnormal grain growth textures in drawn face-centered cubic (fcc) metals

2.1. Silver wire

The annealing textures of cold drawn silver wire were studied by Shin et al. [3]. The Rex textures of cold drawn silver wires were discussed in [1]. When annealed for 1 min at 600°C, a 99.99% pure silver wire cold-drawn by 99% reduction in area was almost completely recrystallized (Rexed). The Rex texture was composed of the <100>//AD component in the majority and the <111>//AD component in the minority, where AD stands for the axial direction. However, as annealing time was prolonged, the <111>//AD component became higher than the <100>//AD component, and the orientation density ratio of <111>//AD to <100>//AD increases, accompanied by abnormal grain growth.

The texture change during grain growth after Rex must be related to grain boundary characteristics. In this stage, dislocation cannot influence the texture change because of its low density. The abnormal grain growth texture was discussed with reference to a theory developed by Abbruzzese and Lücke [4]. This theory allows to quantitatively take into account the influence of textures on grain growth. The theory is based on the statistical model of grain growth proposed by Hillert [5] and modified by Abbruzzese [6]. Shin et al. [3] considered a system composed of two major orientations, *A* and *B*, and other minor orientations, *C*. For the system, the growth rates of the *A*-, *B*-, and *C*-oriented grains of the size class *i* are expressed as [3]

$$\frac{dR_i^A}{dt} = \tilde{M}^A \left(\frac{1}{\tilde{R}_C^A} - \frac{1}{R_i^A} \right) \quad (1)$$

$$\frac{dR_i^B}{dt} = \tilde{M}^B \left(\frac{1}{\tilde{R}_C^B} - \frac{1}{R_i^B} \right) \quad (2)$$

$$\frac{dR_i^C}{dt} = \tilde{M}^C \left(\frac{1}{\tilde{R}_C^C} - \frac{1}{R_i^C} \right) \quad (3)$$

Here

$$\tilde{M}^A = \frac{\Phi^A \overline{R^2}^A M^{AA} + \Phi^B \overline{R^2}^B M^{AB} + \sum_C \Phi^C \overline{R^2}^C M^{AC}}{\Phi^A \overline{R^2}^A + \Phi^B \overline{R^2}^B + \sum_C \Phi^C \overline{R^2}^C} \quad (4)$$

$$\tilde{M}^B = \frac{\Phi^A \overline{R^2}^A M^{AB} + \Phi^B \overline{R^2}^B M^{BB} + \sum_C \Phi^C \overline{R^2}^C M^{BC}}{\Phi^A \overline{R^2}^A + \Phi^B \overline{R^2}^B + \sum_C \Phi^C \overline{R^2}^C} \quad (5)$$

$$\tilde{R}_C^A = \frac{\Phi^A \overline{R^2}^A M^{AA} + \Phi^B \overline{R^2}^B M^{AB} + \sum_C \Phi^C \overline{R^2}^C M^{AC}}{\Phi^A M^{AA} \overline{R}^A + \Phi^B M^{AB} \overline{R}^B + \sum_C \Phi^C M^{AC} \overline{R}^C} \quad (6)$$

http://dx.doi.org/10.5772/59724

$$\tilde{R}_C^B = \frac{\Phi^A \overline{R^2}^A M^{BA} + \Phi^B \overline{R^2}^B M^{BB} + \sum_C \Phi^C \overline{R^2}^C M^{BC}}{\Phi^A M^{BA} \overline{R}^A + \Phi^B M^{BB} \overline{R}^B + \sum_C \Phi^C M^{BC} \overline{R}^C} \tag{7}$$

$\bar{R}$ denotes the mean radius of all the grains and $\bar{R}^H$ that of the grains of texture component H. The different averages can be written as

$$\bar{R} = \sum_i \Phi_i R_i;\ \bar{R}^H = \sum_i \Phi_i^H R_i^H / \Phi^H \tag{8}$$

$$\overline{R^2} = \bar{R}^2 + \sigma^2 = \bar{R}^2(1+\kappa^2) \tag{9}$$

$$\overline{R^2}^H = \bar{R}^{H2} + \sigma^{H2} = \bar{R}^{H2}(1+\kappa^{H2}) \tag{10}$$

with $\Phi_i = \Phi_i^A + \Phi_i^B + \Phi_i^C$; $\Phi^A = \sum_i \Phi_i^A$; $\Phi^B = \sum_i \Phi_i^B$; $\Phi^C = \sum_i \Phi_i^C$ and $\Phi^A + \Phi^B + \Phi^C = 1$.

Φ_i^H denotes the number of grains per unit volume of class (i, H), n_i^H, divided by the total number of grains N_G, i.e.,

$\Phi_i^H = n_i^H / N_G$ with $\sum_H \sum_i n_i^H = N_G$;

$$\sum_H \sum_i \Phi_i^H = 1 \tag{11}$$

σ and σ^H are the standard deviations and $\kappa = \sigma / \bar{R}$ and $\kappa^H = \sigma^H / \bar{R}^H$ the variation coefficients. M^{AB}, the grain growth diffusivity of a boundary between grains of the orientations A and B, is defined as

$$M^{AB} = 2m^{AB}\gamma^{AB} \tag{12}$$

where m^{AB} and γ^{AB} are the mobility and the tension of the boundary between grains of the orientations A and B. It is usually assumed that $M^{AB} = M^{BA}$. According to Eq. 1, $\widetilde{M}^H$ and $\tilde{R}_C^H$, which denote integrated diffusivity and integrated critical radius for the texture component H, respectively, control the growth of component H. $\tilde{R}_C^H$ is the radius dividing the grain sizes of component H into those which grow and those which shrink, and $\widetilde{M}^H$ determines the rate of these processes. Furthermore, a total critical radius R_C and the partial critical radii R_C^H are given by [3]

$$R_C = \overline{R^2} / \bar{R}; R_C^H = \overline{R^2}^H / \bar{R}^H \qquad (13)$$

R_C corresponds to the critical radius of the grain size distribution if all M^{HK} would be equal (then all grains with $R_i > R_C$ would grow and those with $R_i < R_C$ would shrink), and R_C^H would be the critical radius if only the grains of the orientation H would exist [7]. If $M^{AA} > M^{ij}$ with $i \neq A$ and $j \neq A$, $\widetilde{M}^A > \widetilde{M}^B$ and $\widetilde{M}^A > \widetilde{M}^C$. Therefore, the grain growth rate in cluster of *A* grains is the highest, leading to elimination of the *A–A* boundary and thus an increase in size of the *A* grains. The resultant size advantage of *A* grains will cause the growth of *A* at the expense of neighboring *B* or *C* grains [3].

The recrystallized fcc metal wire has a fiber texture composed of major <100>//AD, minor <111>//AD, and others. The grain boundaries of wire can be approximated by tilt boundaries of <100>//AD, <111>//AD, and other grains, along with boundaries between them as shown in Figure 1. If the <111>//AD grain is denoted as *A*, the <100>//AD grains as *B*, and others as *C*, then the above discussion can be applied to the present system. In order for this theory to be applied, grain boundary mobility should be known. To the best of our knowledge, the grain boundary mobility data for silver are not available. However, the mobility data for aluminum indicate that the average mobility of <111> tilt boundary is much higher than that of <100> boundary at elevate temperatures above 400°C (Figure 2). If this tendency holds for the present system, since grain boundary energies are expected to be approximately same in average (Figures 3), *A* grains will grow at the expense of neighboring *B* or *C* grains, as already explained [3].

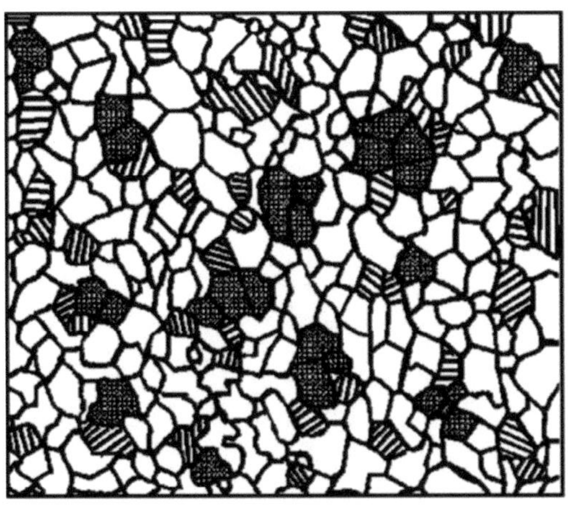

Figure 1. Schematic microstructure of section normal to wire axis, which consists of <100>//AD grains (white), <111>//AD grains (crosshatched), and grains having other orientations (hatched) [3].

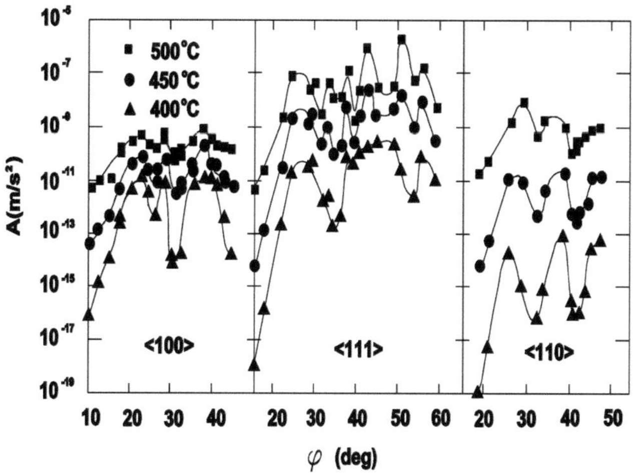

Figure 2. Misorientation dependence of tilt grain boundary mobility in aluminum of misorientation axes <100>, <110>, <111> [8].

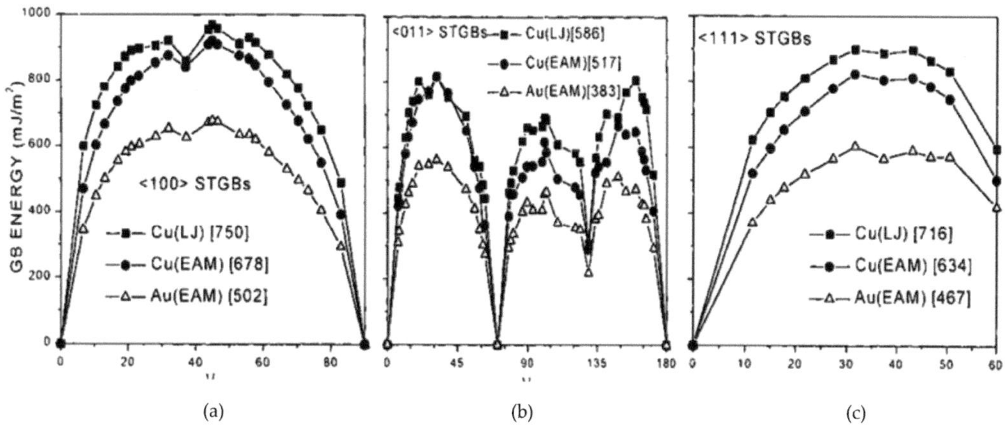

Figure 3. Calculated energy (in mJ/m²) for symmetrical tilt grain boundaries of fcc metals with interface planes perpendicular to (a) <100>, (b) <110>, (c) <111> as a function of tilt angle. LJ and EAM indicate energies calculated using LJ and EAM potentials [9]. Quantity in brackets indicates average tilt boundary energy [3].

2.2. Copper wire

The global texture of drawn copper wires are characterized by the major <111>//AD + minor <100>//AD duplex fiber texture [10-13]. Annealing of the drawn copper wires at low temperatures leads to the development of the major <100>//AD + minor <111>//AD duplex fiber Rex texture [11-13]. Park and Lee [13] studied annealing textures of 90% drawn copper wires. The global deformation texture consisting of major <111>//AD + minor <100>//AD orientations changed to the major <100>//AD + minor <111>//AD texture, as shown in Figure 4, which shows the orientation density ratio of <100>//AD to <111>//AD in the drawn copper wire as a function of annealing time at 700°C. The ratio increases steeply up to about 1.8 after annealing for 180 s, wherefrom it decreases and approach about 0.3 after 6 h [13]. The increase in the ratio indicates the occurrence of primary recrystallization [13] and the decrease demonstrates a change in texture during subsequent abnormal grain growth as can be seen in Figure 5. The figure shows optical microstructures of the drawn wires annealed at 700°C for times varying from 10 min to 6 h [13]. Abnormal grain growth takes place first near the surface and the grains grow into the center [13].

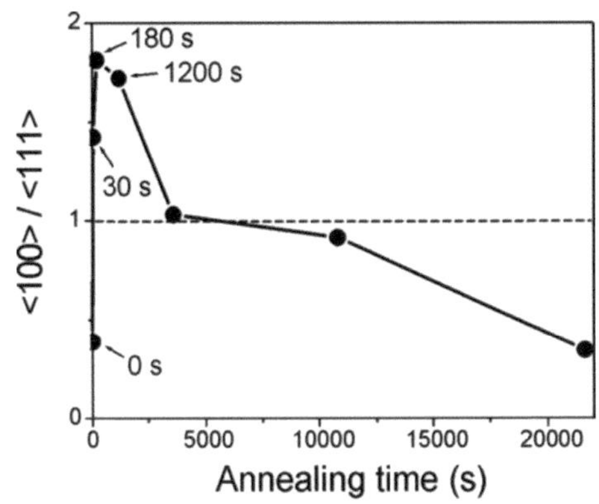

Figure 4. Orientation density ratio of <100>//AD to <111>//AD of 90% drawn copper wires as a function of annealing time at 700°C [13].

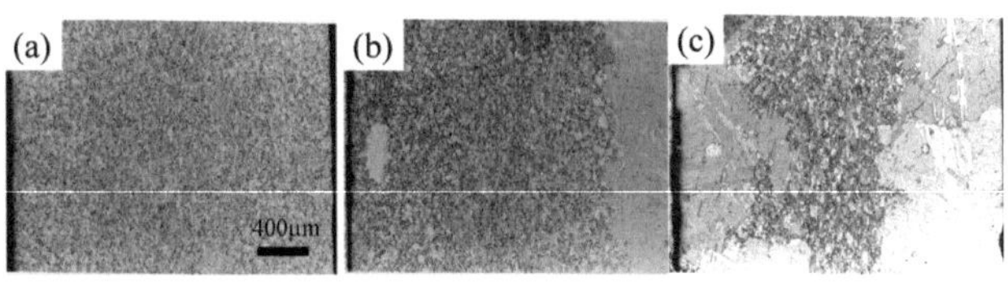

Figure 5. Optical microstructures of copper wires drawn by 90% and annealed at 700°C for (a) 10 min, (b) 1 h, (c) 6 h [13].

The grain size distribution along the radial direction of the longitudinal section of the drawn wire after annealing for 1 h at 700°C shows no considerable abnormal grain growth as shown in Figure 5. The average grain sizes and the size distributions of grains smaller than 2 $\bar{R}$, where $\bar{R}$ stands for the average grain size, were similar in all regions [13]. Relatively large grains in the surface and middle regions had the <111>-<112> and <100>//AD orientations, respectively, as shown in Figures 6 and 7 [13]. The size spread of grains in the middle region is wider than that in the surface region [13]. The center region shows elongated grains, indicating that it is not fully recrystallized [13].

The texture distribution of the copper wire drawn by 90% and annealed for times ranging from 30 s to 1 h at 700°C showed that, for the wire annealed for 30 s, the <112>//AD component was distinct in the center region, although it becomes slightly weaker than the <100>//AD component, while the major <100> + minor <111>//AD texture evolved in the middle region and the major <111> + minor <100>//AD texture in the surface region [13]. The intensity of the <100>//AD orientation was strongest in the middle region. For the wire annealed for 1 h, the major component of the texture of the surface region could be approximated by the {110}<001> orientation (<110>//RD and <001>//AD, where RD stands for the radial direction). The texture of the wire grown abnormally after annealing at 700°C for 1 h could be approximated by a <111>-<112>//AD with a weak <112>//AD component.

As mentioned in Section 2.1, two texture transitions occurred in the 90% drawn copper wires during annealing at 700°C, as shown in Figure 4 [13]. One is the transition from the <111>//AD deformation texture to the <100>//AD Rex texture in the early stage of annealing and the other is the transition from the <100>//AD Rex texture to the <111>-<112>//AD abnormal grain growth texture in the later stage of annealing, neglecting minor components [13]. The later stage transition is now discussed. Unlike the early stage transition, the second stage transition is likely to occur as a result of grain growth driven by grain boundary energy [13]. The microstructural change of the drawn copper wires during annealing at 700°C indicates that the increase of the <111>-<112>//AD components and the decrease of the <100>//AD component are caused by abnormal grain growth (Figure 5) [13]. It can be seen that abnormal grain growth starts to occur just below the wire surface, followed by growth into the center [13]. This result indicates that the surface region is in better position to undergo abnormal grain growth. The abnormal grain growth consistently took place in annealing atmospheres of air, argon, hydrogen, and vacuum (1x10^{-4} Torr) [13]. Therefore, the abnormal grain growth is not related to the wire surface. Typically abnormal grain growth in bulk materials can take place if one or more of the following conditions are satisfied: 1) the presence of dispersion of coarsening pinning particles, 2) a wide spread in grain sizes, and 3) a strong texture [12]. Any second phase particles was not seen in TEM of the annealed specimen. Therefore, the effect of dispersions can be neglected [13]. The grain size distribution along the radial direction of the longitudinal section of the wire drawn and annealed for 1 h at 700°C did not show substantial abnormal grain growth [13]. The average grain sizes and the size distributions of grains smaller than 2 $\bar{R}$, where $\bar{R}$ denotes the average grain size, were similar in all regions [13]. Relatively large grains in the surface, middle, and center regions had the <111>-<112>, <100>, and <111>-<112>//AD orientations, respectively [13]. The size spread of grains in the middle region was the widest, while in the surface region it is the narrowest [13]. According to Hillert's statistical model for grain growth in the steady state, the grain growth rate is given by [5]

$$\frac{dR}{dt} = A\left(\frac{1}{\bar{R}} - \frac{1}{R}\right) \tag{14}$$

where A is a factor depending on mobility, grain boundary energy, and grain shape. The model indicates that a grain having $R < \bar{R}$ will shrink and one with $R > \bar{R}$ will grow; if a grain larger than 2 $\bar{R}$ existed, then abnormal grain growth would occur [13]. Also, in terms of the local size distribution, the larger a grain is in comparison with the neighboring grains, the faster the grain grows [13]. If the size effect controls the texture, the <100>//AD oriented grains will be in better position to grow, making abnormal grain growth occur in the middle region [13]. This prediction is at variance with the experimental results. Therefore, the grain size effect does not seem to be related to the abnormal grain growth accompanying the development of the <111>-<112>//AD texture in the surface region [13]. In order to investigate the effect of texture on abnormal grain growth, the orientations of abnormally growing grains were measured and compared with the local texture of normally growing grains in the near-surface region [13].

Figure 6 shows the texture distribution of the drawn copper wire annealed for 1 h at 700°C. In the annealing condition, extensive abnormal grain growth is not seen (Figure 5 b). The major component of the texture of the surface region can be approximated by <001>//AD; <110>//RD (Figure 6). Special boundaries originating from the evolution of texture is known to play a substantial role in grain growth [8,14-21]. In order to reduce the energy stored in the material in the form of grain boundaries, grain boundaries migrate and the migration of grain boundaries plays an important role in the microstructure and texture evolutions during normal or abnormal grain growth [13]. It has also been reported that the motion of grain boundaries is controlled by the grain boundary mobility at relatively high temperatures, whereas the grain boundary migration is determined by the triple junction mobility at low temperatures [14]. The grain boundary migration depends on the structure and energy of grain boundaries. The migration velocity, v, is generally assumed to be directly proportional to the driving force, P, originated from the grain boundary energy and the mobility, M [15]. That is,

$$v = MP \tag{15}$$

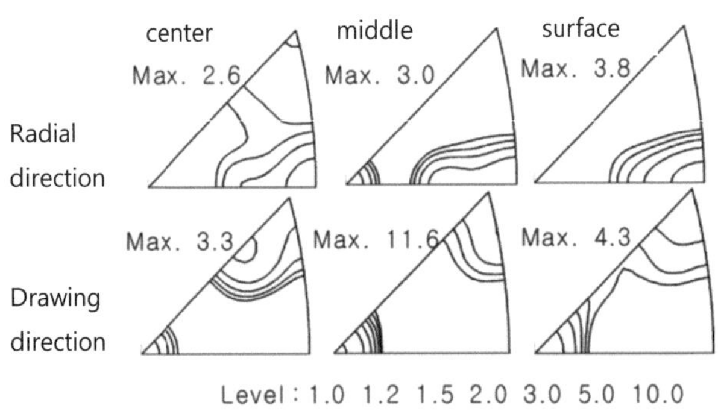

Figure 6. Textures in center, middle, and surface regions of 90% drawn copper wire after annealing for 1 h at 700°C [16].

The grain boundary energy and mobility depend on the grain boundary structure, which can be described by misorientation, causing the growth rate anisotropy. The formation of the special boundaries around a specific grain in a textured matrix has a great effect on the selective grain growth [16-18]. The grain boundary energy and mobility also depend on the segregation of impurity atoms [19]. The grain boundary energy may vary by 20% or less, but the mobility may vary by orders of magnitude [20]. It has been also known that for fcc metals tilt boundaries migrate much faster than the twist boundaries [14]. As mentioned above, the normally growing grains in the surface region of the wire had a fiber texture consisting of major <100>//AD and minor <111>//AD components and the grain boundaries of the wire can be approximated by tilt boundaries of <100> and <111> rotation axes [13]. Grain boundary mobility data for copper are not available, but the data of the misorientation dependence of the tilt grain boundary mobility in aluminum is shown in Figure 2, which illustrates that the average mobility of <111> tilt boundary is far higher than that of <100> boundary above 400°C. If the grain boundary characteristics of copper are similar to those of aluminum, then the mobility of <111> tilt boundaries of copper will be much higher than that of <100> tilt boundaries, and so neighboring the <111>//AD grains will grow faster than neighboring <100>//AD grains, resulting in larger <111>//AD grains, which in turn will grow at the expense of <100>//AD grains because they now have a size advantage [13].

Taking into account that no abnormal grain growth occurred in the middle region whose texture also consists of major <100>//AD and minor <111>//AD components having only a different intensity ratio compared to the surface region [13], it is clear that a statistical condition of the neighboring events of <111>//AD or <100>//AD grains is needed for the abnormal grain growth of <111>//AD grains. The grain size distribution should also be taken into account. In the surface region, the <111>//AD grains larger than the <100>//AD grains will have greater size advantage due to the neighboring event, while in the middle region the <100>//AD grains larger than the <111>//AD grains might be attributed to hindering the growth of <111>//AD grains [13]. The development of the <111>//AD texture at the expense of <100>//AD texture by abnormal grain growth was also observed in drawn silver wire (Section 2.1).

Note that the development of the {110}<001> texture is extensive in the surface region during annealing at 700°C [13]. In the early stage of annealing at 700°C, the texture is characterized by a fiber texture, which means the orientations of grains are distributed with equal probability around the wire axis, resulting in weak texture along the radial direction [13]. Further annealing at 700°C leads to the evolution of the textures along the radial direction, and the increase in the intensity is more noticeable in the surface region than elsewhere [13]. The further annealing also leads to the increase in the intensity of the <100>//AD texture in the surface region, compared with that of primarily recrystallized wire. Consequently, the {110}<001> texture developed in the surface region [13]. It is interesting to note that in the surface region the texture resulted from grain growth is identical to the texture formed by dynamic recrystallization during drawing. Even so, the reason of the texture recurrence is ambiguous [13].

Figure 7 shows the orientations of abnormally grown grains located in the surface region of 90% drawn copper wire after annealing for 1 h at 700°C. The orientations of abnormally grown grains are approximated by {110}<112>, and the texture of the neighboring grains by {110}<001> in the surface region [13]. The measurements of the orientations of an abnormally grown grain and of its neighboring grains along both the radial and drawing directions can determine the crystallographic characteristics of the grain boundaries for these two kinds of grains [13]. There has been no prior research that has reported the crystallographic relationship of the special boundaries for abnormal grain growth in wire materials because the texture along the radial

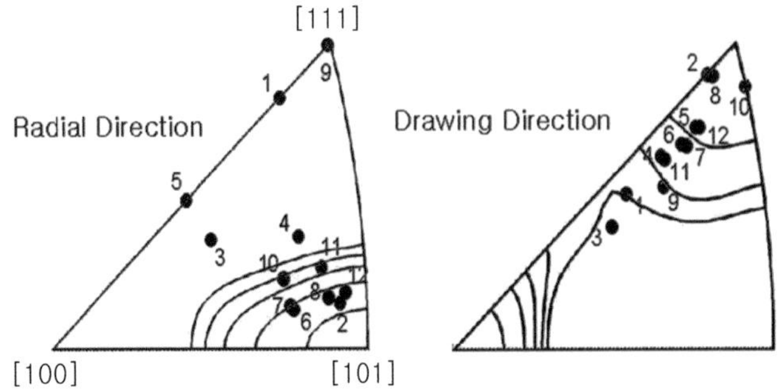

Figure 7. Textures of the surface region (contours) and orientations (black symbols) of large grains (50-250 μm) in the surface region of 90% drawn copper wire after annealing at 700°C for 1 h [13].

direction has not been considered [13]. As can be seen in Figure 8, the {110}<112> orientation of abnormally grown grains and the {110}<001> orientation of neighboring grains, for example, (1-1 0)[112] and (1-1 0)[001], can form a 35.3°<110> tilt grain boundary [13]. The misorientation dependence of the tilt grain boundary mobility in aluminum in Figure 2 shows that 35°<110> tilt boundary leads to a high mobility at high temperatures [13]. These indicate that the grain boundaries formed by the {110}<112> and {110}<001> grains in the surface region of the copper wire have higher migration rates than other boundaries, resulting in the abnormal grain growth [13]. Twinning did not seem to play a role in the progress of the abnormal grain growth in the copper wire from the observation that coherent twin boundaries did not form boundaries between the abnormally grown grains and the matrix, and even isolated twins were found inside the abnormally grown grains [13].

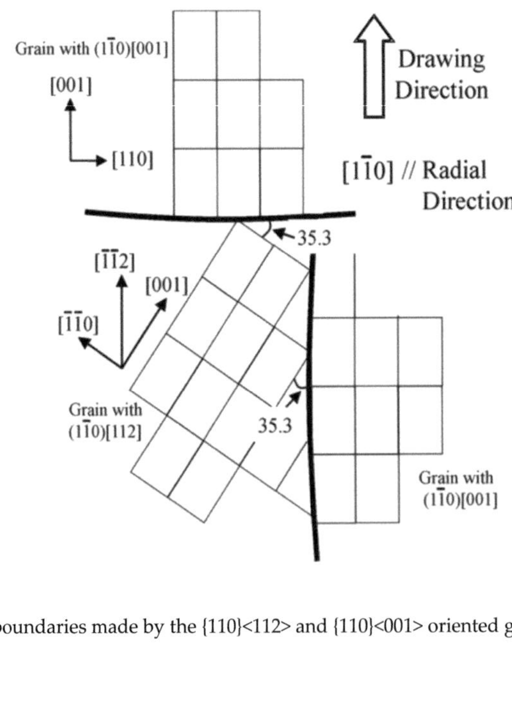

Figure 8. The 35.3°<110> tilt boundaries made by the {110}<112> and {110}<001> oriented grains in fcc lattice [13].

3. Abnormal grain growth textures in nanocrystalline deposits

3.1. Freestanding nanocrystalline Ni and Fe-Ni electrodeposits

The texture change during annealing of a Fe-Ni electrodeposit of several nanometers in grain size was reported by Czerwinski et al. [21]. The Fe-Ni electrodeposit had initially a duplex texture of <100>//ND and <111>//ND with the pole density of the former being a little higher than that of the latter, where ND indicates the deposit-surface normal direction. As annealing proceeded, the pole density of <111>//ND rapidly increased as shown in Figure 9. Similar results were obtained by Park et al. [22,23]. They obtained 30-μm-thick Ni and 20-μm-thick Ni-20 % Fe electrodeposits from Watts-type solutions. The grain size of the Ni electrodeposit was estimated to be 15 to 25 nm, and the grain size of the Ni-20 % Fe deposit was estimated to be 10 to 20 nm. The textures of the both deposits were approximated by the major <100>//ND + minor <111>//ND, which changed to the major <111>//ND + minor <100>//ND after annealing as shown in Figures 10 and 11. Figure 10 shows {111} and {100} pole figures of the nanocrystalline Ni electrodeposit before and after annealing at 300°C for 5 min. Figure 11 shows {111} and {100} pole figures of the nanocrystalline Ni-20% Fe electrodeposit before and after annealing at 400°C for 30min. The pole figures were measured with an X-ray pole figure device (Target: CoK_α). These results indicate that the deposition texture characterized by major <100>// ND and minor <111>//ND components changes to the texture consisting of major <111>//ND and minor <100>//ND components after annealing. It should be noted that all the deposits are free standing foils separated from the substrates and the grain sizes of the foils are in the order of 10 nm. Such small grains have little dislocations and so the dislocation effect on the texture transition can be excluded. The thicknesses of the foils are over 1000 times larger than the grain size. Therefore, the surface energy effect can be neglected. The most important factors dominating the texture change are the grain boundary energy and mobility. In this case, the texture transition is similar to that in abnormal grain growth textures in drawn fcc metals (Section 2). The explanation in Section 2 can apply to the present case [24,25]. Park et al. [22,23] also found that the lattice constants of the <100>//ND oriented grains in the as-deposited Ni and Ni-Fe alloy were larger than those of the <111>//ND oriented grains. Lee [26] suggested one explanation of the lattice constant problem based on differences in concentration between the <100>//ND and <111>//ND grains. It should be mentioned that all the data were obtained from freestanding samples. Therefore, any stress effect due to the substrate is excluded. Organic additives in electroplating baths are believed to be partly adsorbed on the deposit, some of which are likely to decompose under high electric fields during electrolysis [26]. Whatever the adsorbates may be, they can hinder growth of electrocrystallites, which in turn results in nanocrystalline deposits [26]. Figure 12 (a) shows that the (111) and (200) interplanar spacings after annealing are on the extrapolated lines of the data at 400°C and 500°C, respectively. The extrapolated lines are parallel to each other [26]. The thermal expansion coefficient obtained from the lines is $15.46 \times 10^{-6}\ K^{-1}$, which is a little higher than the reported value ($12.3 \times 10^{-6}\ K^{-1}$) [27,28].

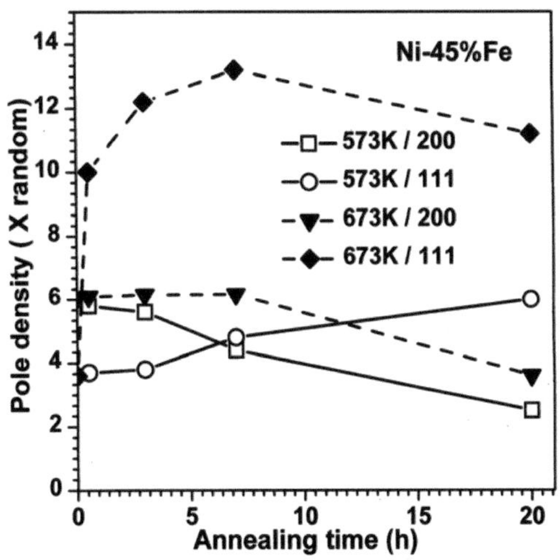

Figure 9. Change in texture of nanocrystalline Fe-Ni electrodeposit during annealing at 573 and 673 K [21].

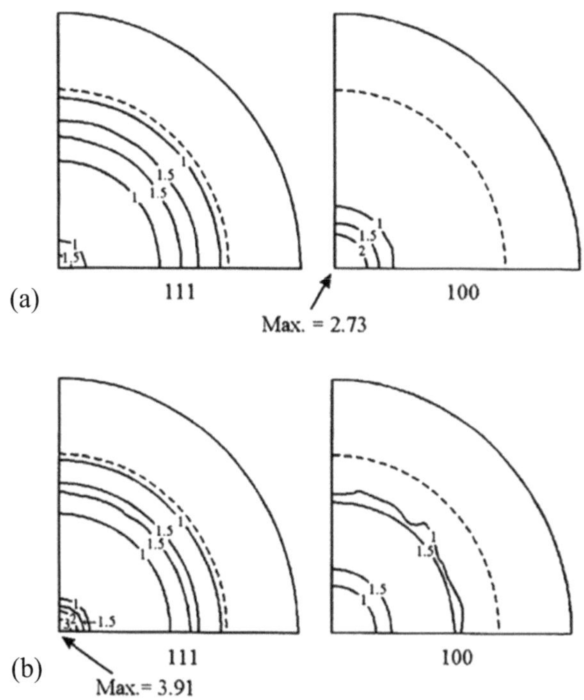

Figure 10. {111} and {100} pole figures of Ni electrodeposit: (a) before and (b) after annealing for 5 min at 300°C [23].

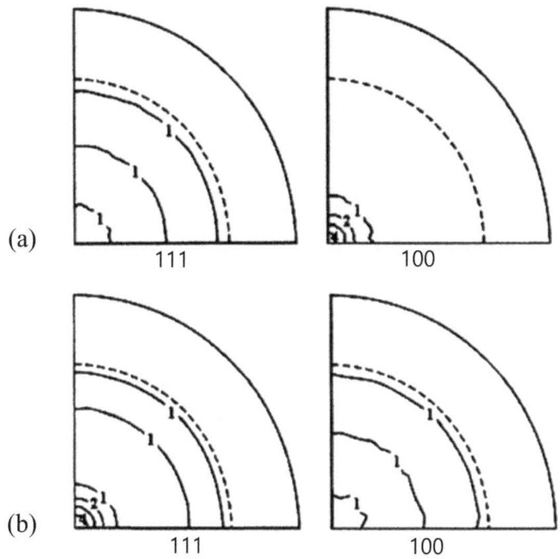

Figure 11. {111} and {200} pole figures of 20%Fe-Ni electrodeposit: (a) before and (b) after annealing for 30 min at 400°C [22].

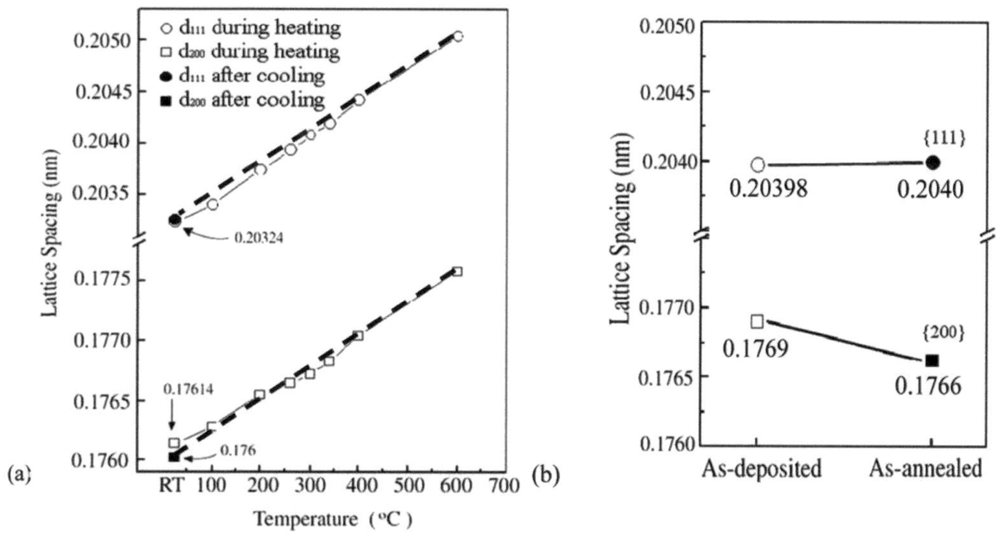

Figure 12. (200) interplanar spacing of <100>//ND grains and (111) interplanar spacing of <111>//ND grains in (a) Ni electrodeposit as a function of temperature and in (b) Fe-80%Ni electrodeposit before and after annealing for 30 min at 400°C [23].

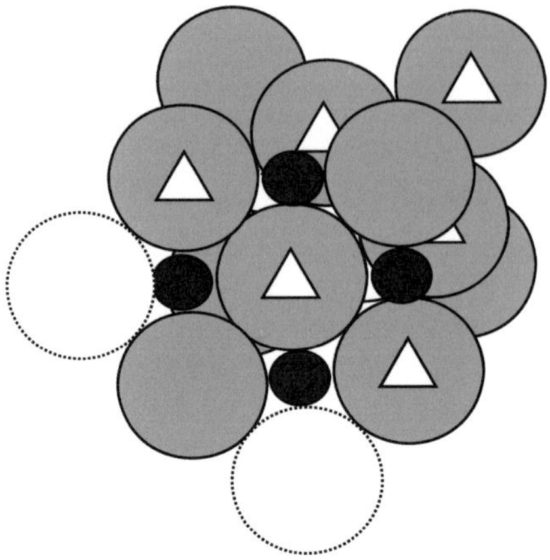

Figure 13. Fcc unit cell showing largest interstitial holes (black spheres) on one (100) plane. Plane made of spheres marked with white triangles is (111) plane. Radius of interstitial hole = $(2_{1/2}-1)$ (atom radius) [26].

The fact that the (111) interplanar spacing at room temperature changes very little after annealing, while a decrease in the (200) interplanar spacing at room temperature after annealing implies that adsorption of something increasing the lattice constant on the {100} planes normal to the growth direction of deposits (ND) is preferred to adsorption on the {111} planes normal to ND [26]. The largest interstitial holes in the fcc lattice are located at <1/2 0 0> as shown in Figure 13. Therefore, adsorption on the {100} planes of the <100>//ND grains is expected to be favored than on the {111} planes of the <111>//ND grains [26]. One candidate of products from decomposition of the organic additives can be carbon [26]. If carbon atoms occupy the <1/2 0 0> interstitial holes, the lattice constants increase and the interplanar spacing increases because the interatomic distance of carbon (graphite) is 0.1421 nm (Table 1), which is larger than the diameter of the largest interstitial hole of nickel, $(\sqrt{2}-1)\,D_{Ni} = 0.1032$ nm, where D_{Ni} is the interatomic distance of Ni (0.2492 nm) [26]. However, when the deposit undergoes grain growth during annealing, the interstitial atoms in the <100>//ND grains are likely to be ejected to grain boundaries and some of them diffuse to the <111>//ND grains [26]. In this way, the lattice constants of the <100>//ND grains decrease and those of the <111>//ND grains slightly increase. This seems to be the case of the result in Figure 12 (a) [26]. The interatomic distance of fcc Fe at 25°C was calculated to be 0.2579 nm $(1+23\times10^{-6}(25-916))$ = 0.2526 nm (Table 1, Calc.) [26].

Constant	Ni	fcc Fe (911 to 1392°C)	Graphite	Ref.
Atomic weight	58.71	55.85	0.1421 nm	[29]
Mean therm. expans. coeff.	-	$23\times10^{-6}K^{-1}$(916 to 1388°C)		[30]
Interatomic distance	0.2492 nm	0.2579 nm (916°C)		[29]
		0.2526 nm (25°C)		Calc.

Table 1. Physical constants of Ni, fcc Fe and C

20 % Fe-80 %Ni = 20.8 atomic % Fe-79.2 atomic% Ni
(111) interplanar spacing of Ni = interatomic distance × $(2/3)^{1/2}$ = 0.20347 nm
(200) interplanar spacing of Ni = interatomic distance × $2^{-1/2}$ = 0.1762 nm
(111) interplanar spacing of Fe = interatomic distance × $(2/3)^{1/2}$ = 0.20625 nm
(200) interplanar spacing of Fe = interatomic distance × $2^{-1/2}$ = 0.1786 nm
(111) interplanar spacing of Fe-Ni = 0.20625 nm × 0.208 +0.20347 nm × 0.792 = 0.2040 nm
(200) interplanar spacing of Fe-Ni = 0.1786 nm × 0.208 + 0.1762 nm × 0.792 = 0.1767 nm

Table 2. Lattice spacing at 25°C calculated using data in Table 1 [26]

The (200) and (111) interplanar spacings calculated using the interatomic distance of Ni (0.2492 nm) are 0.1755 nm and 0.2027 nm, respectively [26]. The (200) interplanar spacing of the annealed <100>//ND grains calculated using the synchrotron XRD data (0.176 nm in Figure 12(a)) is 1.0028 times larger than 0.1755 nm [26]. The (111) interplanar spacing of the annealed <111>//ND grains calculated using the synchrotron XRD data (0.20324 nm in Figure 12 (a)) is 1.0027 times larger than 0.2027 nm [26]. This indicates that the lattice constants of the annealed <100> and <111>//ND grains are almost the same, and are 1.0027 to 1.0028 times larger than those of pure nickel due to some interstitial elements [26].

For the Fe-Ni deposit, Lee [2] evaluated the composition based on the data in Figure 12 (b). For an fcc crystal, the ratio of the (111) interplanar spacing to the (200) interplanar spacing is $2/\sqrt{3}$ (=1.1547) [26]. The ratio of the (111) interplanar spacing in the <111>//ND grains to the (200) interplanar spacing in the <100>//ND grains of the annealed specimen is 1.15496 (=1.000225 times the theoretical value) that is almost the same as the theoretical value, while the ratio of the (111) interplanar spacing in the <111>//ND grains to the (200) interplanar spacing in the <100> grains of the as-deposited specimen is 1.15305 that is 1.00143 times smaller than the theoretical value [26]. This result indicates that the composition of the annealed specimen is homogenized, whereas the lattice constants of the <100>//ND grains in the as-deposited specimen are larger than those in the <111>//ND grains [26]. Assuming that the annealed specimen is completely homogenized, the interplanar spacing has been adjusted so that the ratio of the (111) interplanar spacing to the (200) interplanar spacing in Figure 12 (b) is $2/\sqrt{3}$. The adjusted (111) and (200) interplanar spacings are 0.20399 nm and 1.7666 nm, respectively. From the data in Figure 12 (b) and Table 2, we obtain

$$0.20347X_{\mathrm{Ni}} + 0.20625X_{\mathrm{Fe}} = 0.20399 \tag{16}$$

$$\text{or } 0.1762X_{\mathrm{Ni}} + 0.1786X_{\mathrm{Fe}} = 1.7666 \tag{17}$$

$$X_{\mathrm{Ni}} + X_{\mathrm{Fe}} = 1 \tag{18}$$

where X_{Ni} and X_{Fe} are the atomic fraction of Ni and Fe, respectively. Solving the above equations, we obtain X_{Ni} = 0.81 and X_{Fe} = 0.19. These values are equivalent to 18.2 % Fe and 81.8 % Ni, respectively [26].

As in the Ni electrodeposit, the (111) interplanar spacing of the <111>//ND grains (0.203975 nm) in the as-deposited specimen is slightly smaller than that in the annealed specimen (0.20399 nm), while the (200) interplanar spacing (0.1769 nm) of the <100>//ND grains in the as-deposited specimen is 1.00136 times higher than that in the annealed specimen (0.17666 nm) [26]. The value of 1.00136 is higher than that in the Ni specimen (1.0008), and so, the higher value may not be due only to adsorption of interstitial elements [26]. The interatomic distance of fcc Fe is larger than that of Ni (Table 1), and so the larger lattice spacing in the as-deposited specimen can be attributed to higher concentrations of Fe and interstitial element in the <100>// ND grains than in the <111>//ND grains [26]. The adsorption of interstitial elements is preferred on the <100>//ND grains during electrodeposition, as discussed in the Ni deposit, and hence the interatomic distance of the <100>//ND grains is larger than that of the <111>//ND grains, which in turn gives rise to preferred electrodeposition of Fe on the {100} surfaces of the <100>// ND grains and the higher concentration of Fe in the <100>//ND grains [26]. Since the concentration of interstitial elements in the <111>//ND grains is negligible, the difference in the (111) interplanar spacing between before and after annealing can be attributed to differences in the Fe concentration [26]. Let X_{Ni}^{111} and X_{Fe}^{111} ($X_{Ni}^{111} + X_{Fe}^{111} = 1$) be the atomic fractions of Ni and Fe in the <111>//ND grains in the as-deposited specimen, respectively. It follows from the data in Figure 12 (a) and Table 2 that

$$0.20347X_{Ni}^{111} + 0.20625X_{Fe}^{111} = 0.20398 \tag{19}$$

Solving the above equation, we obtain $X_{Ni}^{111} = 0.8165$ and $X_{Fe}^{111} = 0.1835$, which are higher than the mean atomic fraction of Ni ($X_{Ni} = 0.81$) and lower than the mean atomic fraction of Fe (X_{Fe} =0.19). We calculate the atomic fraction of Fe X_{Fe} in the <100>//ND grains in the as-deposited specimen neglecting interstitial elements. From the following equation

$$0.1786X_{\mathrm{Fe}} + 0.1762\left(1 - X_{\mathrm{Fe}}\right) = 0.1769 \tag{20}$$

we obtain $X_{\mathrm{Fe}} = 0.2917$. This is higher than the mean atomic fraction (0.19) by 0.10 or 53%. This value seems to be too high [26]. If we assume that the effect of interstitial elements is similar to that in the Ni electrodeposit, the value of 0.1769 nm should be reduced as

$$0.1769 \text{ nm} \times 1.76/1.7614 = 0.17676 \text{ nm} \tag{21}$$

If we use this value instead of 0.1769 nm, then $0.1786\, X_{\mathrm{Fe}} + 0.1762\,(1 - X_{\mathrm{Fe}}) = 0.17676$ and we obtain $X_{\mathrm{Fe}} = 0.23$. This is 21% higher than the mean atomic fraction [26].

To conclude, for the nanocrystalline Ni electrodeposits, the <100>//ND grains contains higher carbon atoms located in their interstitial sites than the <111>//ND grains, and for the nanocrystalline Ni-Fe electrodeposit, the concentration of carbon and iron in the <100>//ND grains is higher than that in the <111>//ND grains [26]. There has been no single report in which the concentration of grains varies with their orientation in a given electrodeposit. Direct experimental verification of the present hypothesis is desirable. The texture changes during annealing in Figures 9 to 11 are in variance with the texture changes in electrodeposits during Rerecrystallization [1,31], in which both the <100>//ND and <111>//ND deposition textures change to the <100>//ND texture after Rex., for fcc metallic deposits. Therefore, we expect that the <100>//ND+<111>//ND duplex deposition texture changes to the <100>//ND texture after annealing. However, the texture changes in Figures 9 to 11 are opposite to our expectation. The annealing texture includes the recrystallization texture and the abnormal grain growth texture [2]. When the deposit is thick enough to neglect the surface energy and/or the interface energy between the deposit and substrate compared with the internal energy due to dislocations, the deposition texture may change after recrystallization [1]. For fcc metals, the <100>, <110>, and <111>//ND deposition textures change to the <100>, <$\sqrt{3}$10>, and <100>//ND textures after recrystallization, respectively, as explained in [1,31]. The evolution of recrystallization textures are well explained by the strain-energy-release maximization (SERM) model [1]. If deposits are obtained at high temperatures or if the grain size of deposits is smaller than the order of 10 nm, the deposits may not have high enough dislocation density to cause recrystallization, and their annealing textures cannot be explained by the SERM model due to dislocations [1]. In this case, the annealing textures will be controlled by stresses that are not caused by dislocations, grain boundary characteristics, and surface and interface energies [2]. When the stresses are dominant, the annealing texture is determined such that the strain energy of deposit is minimal [2]. The texture changes in deposits under the plane-stress, equibiaxial-strain state are discussed in Section 3.2. The plane-stress, equibiaxial-strain state of deposits is caused by their substrates.

Now we are in position to discuss the texture changes in Figs. 9 to 11, the duplex texture of major <100>//ND + minor <111>//ND changing to the texture of major <111>//ND + minor <100>//ND after annealing. The grain sizes of the deposits are in the order of 10 nm. Therefore, the dislocation effect on the texture change can be excluded. The deposits are separated from substrates, and the thickness of them is more than 1000 times larger than the grain size. Therefore, the effects of the interface and surface energies and the strain energy due to different thermal expansion coefficients of deposit and substrate can also be neglected.

It is apparent that the most important factors dominating the texture change are the grain boundary energy and mobility. Lee and Hur [24,25] explained the texture turnover from a (<111>//ND + <100>//ND) to the <111> //ND in fcc metals. The grain boundaries in a material with the duplex fiber texture of <111>//ND + <100>//ND may approximate to tilt boundaries of <111> and <100> //ND grains and boundaries between <111>//ND and <100>//ND grains as shown in Figure 1 [25]. This problem is the same as the texture turnover in Sections 2.1 and 2.2. Therefore, the discussion in Sections 2.1 and 2.2 applies to the present case. In addition to the explanation in Sections 2.1 and 2.2, the <111>//ND grains in the present deposits may be purer than the <100>//ND grains. Therefore, the <111>//ND grains are likely to have higher grain boundary mobilities than the <100>//ND grains and grow at the expense of the <100>//ND grains.

3.2. Electroless Ni-Co-P alloy deposits

Lee and Hur [24, 25] studied texture changes during annealing of the nanocrystalline electroless Ni-Co-P deposits. Electroless deposition is defined as the formation of metal coatings resulting from autocatalytic chemical reduction of metal ions from solution [31]. The metal being deposited is itself catalytic for continuous chemical reduction of the metal ions, and deposition can proceed solely on the desired surface at a rate essentially linear with time. Electrons for electroless deposition are furnished by the reducing agent, unlike electrodeposition which requires a counterelectrode and electrons supplied via a rectifier. Electroless deposition is always associated with the evolution of substantial amounts of hydrogen gas on the depositing surface, whereas hydrogen gas may or may not be evolved during electrodeposition. Otherwise, electroless deposition can be considered analogous to electrodeposition at constant current density; dense deposits are produced and there is no theoretical limit to the coating thickness that can be formed. The electroless Ni-Co-P alloy films were deposited on a preplated 5086 aluminum alloy sheet [24]. The preplating process included cleaning and nitric acid and double zincating treatments [24]. Oxide film formed after alkali cleaning and acid etching was removed by the zincating treatments, to prevent reoxidation and to form catalytic surface. The composition of the electroless deposition bath was 0.07–0.1 mol/L Ni^{2+} ($NiSO_4$), 0–0.03 mol/L Co^{2+}($CoSO_4$), 0.5 mol/L $(NH_4)_2SO_4$, 0.2 mol/L NaH_2PO_2, 0.2 mol/L sodium citrate, and 1 ppm thiourea [24]. The pH of the bath was adjusted to 12 using ammonia solution. The bath temperature was maintained at 90°C [24]. The thickness of deposits ranged from 30 to 40 μm. The micrographs indicated that fine grains were imbedded in amorphous matrix (Figure 14). The amorphous phase is supposed to have a higher phosphorous content than crystalline grains from the well known fact that the electroless Ni-P deposits tend to become amorphous with increasing P content. The electroless Ni-Co-P alloy deposits showed strong (111) and very weak (200) reflection peaks (Figure 15). Their grain size was estimated to be 6 to 7 nm by the Scherrer equation. When annealed at various temperatures for 2 h, the Ni-Co-P deposits showed that the (200) peak intensity increases more rapidly than the (111) intensity with increasing annealing temperature and cobalt content in the deposits.

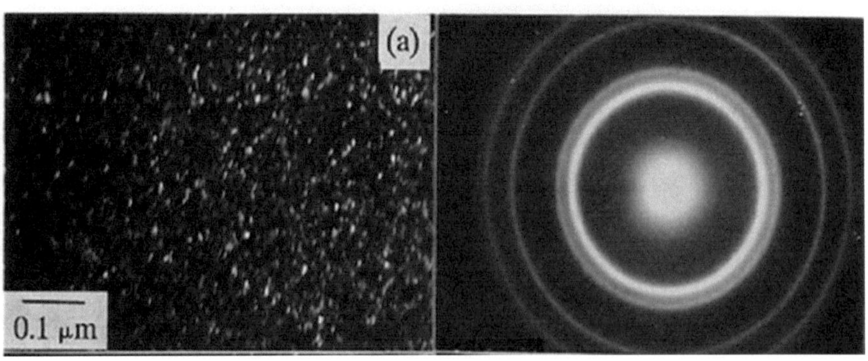

Figure 14. Dark field micrograph and SAD pattern of electroless Ni-2.4Co-7.6P [25].

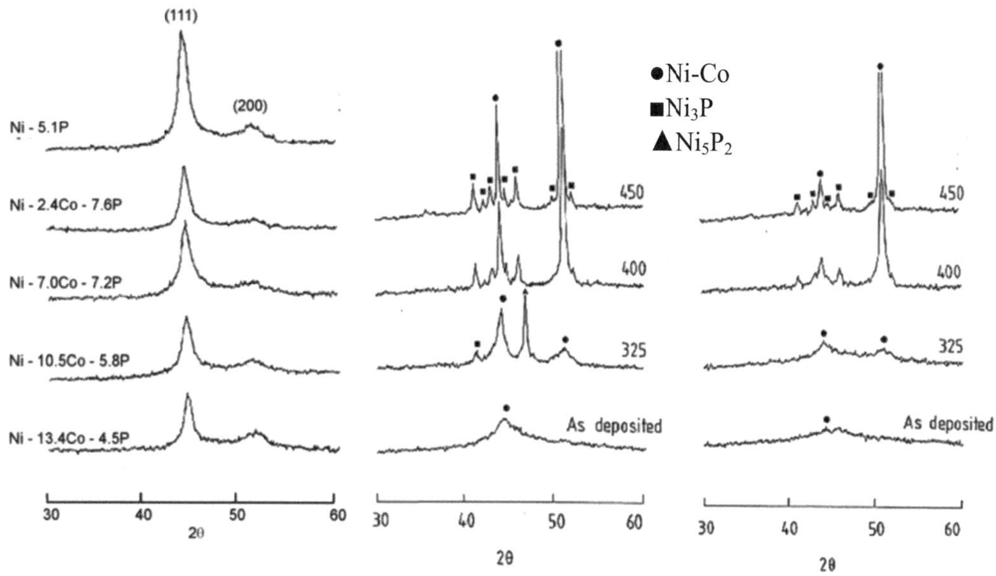

Figure 15. X-ray diffraction patterns of (from left) as-deposited, Ni-2.4Co-7.6P, Ni-10.5Co-5.8P electroless deposits annealed for 2 h. Peak intensities in Ni-2.4Co-7.6P, Ni-10.5Co-5.8P electroless deposits are attenuated [24,25].

At an annealing temperature of 325°C, Ni_5P_2 forms especially in the higher phosphorous deposits, but it disappears and stable Ni_3P appeared at higher temperatures. The similar phenomenon was observed in electroless Ni-Cu-P alloys [31]. The formation of Ni_5P_2 from the annealed Ni-Cu-P deposit is due to a high segregation of P in grain boundary region [25]. When annealed, the high P concentration region was considered to be further increased by extraction of P dissolved in nickel grains, which in turn precipitated the Ni_5P_2 phase. The same explanation can be applicable to the Ni-Co-P deposits [25].

The evolution of the <100>//ND texture in the <111>//ND textured fcc metal films on a substrate during annealing can be attributed to a few sources [32]. The lattice match at the interface between the film and the substrate may lead to the <100>//ND orientation. But the lattice match between the substrate and the nickel alloy deposit has little possibility. When the <111>//ND textured fcc electrodeposits are recrystallized, the <100>//ND texture evolves, as explained in [1,33]. However, this explanation applies to deposits whose dislocation density is high enough to cause recrystallization [33]. Since the grain size of the present deposits is only several nanometers, they are likely to have little dislocation, and so the explanation cannot apply to the present deposits. When grain boundaries control the texture evolution during annealing, the <111>//ND texture is expected to form (Sections 2.1–3.1). The texture change from <111>//ND to <100>//ND may be caused by the fact that the <111>//ND grains have the highest strain energy while the <100>//ND grains the lowest strain energy [34]. The deposit and substrate, which have different thermal expansion coefficients, undergo different volume changes due to a temperature change [24]. In addition, the formation of nickel phosphide could lead to volume changes [24]. These changes can produce strains and in turn stresses in the deposits during annealing [24]. The experimental results indicated that the texture change was more influenced by the Co concentration than by P, implying that the volume changes due to phosphide formation are not a dominant factor leading to the <100>//ND texture [24]. Therefore, this factor is not considered here.

The strains and stresses due to the different thermal expansion coefficients are calculated. The thermal strain ε_{th} due to the different thermal expansions is calculated by [24]

$$\varepsilon_{th} = \int_{T_0}^{T} (\alpha_s - \alpha_f) dT \tag{22}$$

where T and α are temperature and thermal expansion coefficient, respectively. The subscripts 0, s, and f indicate initial state, substrate, and film, respectively. Setting T = 250 to 450 ℃, T_0 = 90 ℃, $\alpha_s = \alpha_{5086Al} = 25.8\times10^{-6}$ K^{-1} [28], $\alpha_f = \alpha_{Ni} = 13.3\times10^{-6}$ K^{-1} [28], we obtain $\varepsilon_{th} = (2\ to\ 4.5)\times10^{-3}$. To be more rigorous, the thermal expansion coefficients of fcc Ni-Co solid solutions should be used. The thermal expansion coefficient of cobalt (hcp) is 13.8×10^{-6} K^{-1} [28], which is very close to that of nickel. Therefore, the thermal expansion coefficient of nickel can be a good approximation of the alloys in the absence of experimental values.

A thin deposit on a thick substrate undergoing isotropic thermal expansion is expected to be under a plane stress, equibiaxial strain state. Cubic-symmetry metals undergo isotropic thermal-expansion. Therefore, the Ni-Co-P deposits are likely to be under a plane stress, equibiaxial strain states. The elastic strain energies per unit volume (the strain energy density) of the <100> and <111>//ND films under an equibiaxial strain ε are given by Eqs. 19 and 20, respectively [35,36].

$$w_{(100)} = \varepsilon^2 \left(S_{11} + S_{12}\right)^{-1} \tag{23}$$

$$w_{(111)} = \left[S_{11} + S_{12} + (1/6)(S_{44} - 2S_{11} + 2S_{12})\right]^{-1} \tag{24}$$

Setting S_{11} = 0.00734, S_{44} = 0.00802, S_{12} = -0.00274 GPa^{-1} at room temperature [37], and ε = $\varepsilon_{th} = (2\ to\ 4.5)\times10^{-3}$, $w_{(100)}$ =0.87 to 4.4 MJ m^{-3}, $w_{(111)}$ =1.55 to 7.86 MJ m^{-3}.

Thus, the <100>//ND grains have lower strain energy than the <111>//ND grains. These values are based on room temperature data [24]. The stresses at experimental temperatures of 250 to 450°C are likely to be lower, but still substantial [24]. To conclude, the evolution of the <100>//ND texture during annealing is caused by preferred growth of <100>//ND grains at the expense of <111>//ND grains under thermal stresses [24].

Electroless Ni-Cu-P deposits show similar grain-size and textures at the initial state [36], but the <111>//ND texture remains relatively stable even after annealing, even though the <100>//

http://dx.doi.org/10.5772/59724

ND texture tends to develop with increasing temperature (Figure 15). This may be due to the fact that the thermal expansion coefficient of copper (19.6×10^{-6} at 700 K [28]) is higher than that of cobalt.

4. Aluminum and copper interconnects

4.1. Textures of Al-1%Cu Interconnects

Field et al. [37,38] deposited 500-nm-thick Al-1%Cu films over Ti and TiN sublayers onto a Si (001) single crystal substrate. The films were patterned into test structures with lines varying from 0.5 to 4 μm in width. A post-patterning anneal at 460°C was then applied to determine its effects on the evolution of texture and grain boundary structure of lines. The resulting structures showed a near-bamboo character for those lines of less than 2.0 μm in width and a polycrystalline structure for the wider lines. The resultant grain size was of the order of 1 μm and varied in proportion to the line width. For an ideal <111>//ND texture, the distribution of crystallographic planes aligned with the grain boundary plane would consist of an arc of uniform intensity extending from the {110} to the {211} in the stereographic unit triangle shown in Figure 16. For wide polycrystalline lines, the measurements of planes along the boundary approximated that of an ideal distribution [2]. For narrow, bamboo-structured lines, the distribution was weighted heavily in favor of the (110) planes aligned with the grain boundary [37,38]. This implies that the line aligns with the <110> direction (Figure 17). The <111>//ND texture strength increased with decreasing line width during the post-patterning anneal [37,38].

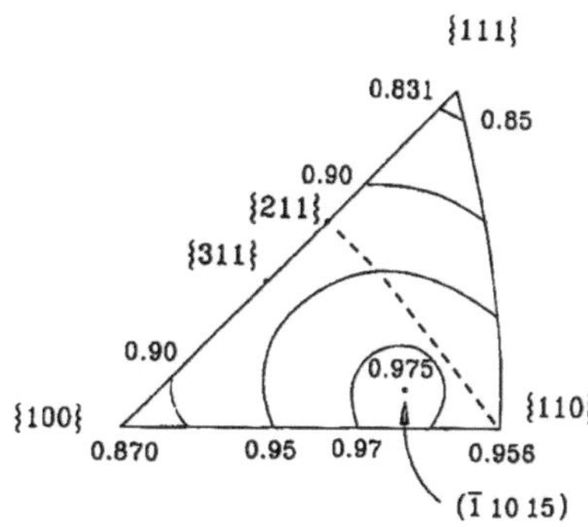

Figure 16. Measured relative surface energy plot for fcc crystal [40]. Dashed line indicates poles of planes normal to {111} planes.

Lee and Lee [2] explained the result in terms of the surface, interface, and strain energies. The surface energy of the (112) plane is the lowest among boundaries normal to the (111) plane (Figure 16) [2]. It is reasonable to assume that the boundaries in the interconnect structure are <111> tilt boundaries [2]. According to the calculated <111> tilt boundary energy (Figure 3c),

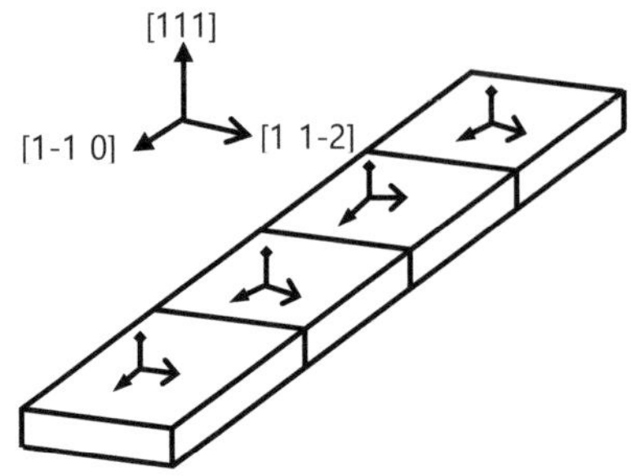

Figure 17. Orientations of grains in bamboo structured line; its texture is approximated by {111}<110> [39].

the (110) boundary does not have the minimum energy, but the boundary energy decreases with decreasing misorientation angle [2]. The line deposits must have had the <111>//ND texture and low dislocation density due to a relatively high deposition temperature, if the usual deposition conditions were used, although the temperature is not given in the papers [38,39]. Therefore, the <111>//ND texture after the post-patterning anneal is believed to result from the grain growth (Section 2 and 3.1). However, the interconnect lines are not guaranteed to have the <111>//ND texture during annealing [2]. Two directional properties should be considered. One is the surface energy and the other is the thermal stress [2]. Among surfaces normal to the (111) surface the {112} planes have the lowest energy (Figure 16). Therefore, when the sidewall surfaces and the surface parallel to the substrate consist of the {112} and {111} planes, respectively, the surface energy of the line has the lowest surface energy. If the surface energy dominates the energy of the line during annealing, the line will be parallel to one of the <110> directions as shown in Figure 17.

4.2. Copper interconnects

The properties of deposited metal films and interconnect structures are sensitive functions of microstructural features. For example, Al-based interconnects (Section 4.1) with a large uniform grain size and a strong <111>//ND texture provide significantly improved reliability. Therefore, an understanding of factors which control microstructural evolution is important for the development and design of reliable, manufacturable interconnect structures, especially in damascene-processed trench interconnects [2].

A structure of damascene trenches is shown in Figure 18. The damascene trenches of various width/space combinations are prepared as grating arrays in a 700-nm-thick FSG layer formed on a Si wafer. A TEOS layer of 300 nm in thickness and subsequently a silicon nitride layer of 55 nm in thickness are vapor-deposited between the FSG and the Si substrate as an etch-stop layer. The trench pattern is made by the lithography on the FSG. TaN of 10 nm in thickness is first sputter-deposited on the FSG as a diffusion barrier and an adhesion layer, followed by sputter-deposition of 150 nm Cu seed layer to serve as cathode for electroplating. The trenches are filled with Cu by electroplating in an acidic sulfate solution with commercial additives. The thickness of Cu deposit over the trench grating is about 700 nm [41].

http://dx.doi.org/10.5772/59724

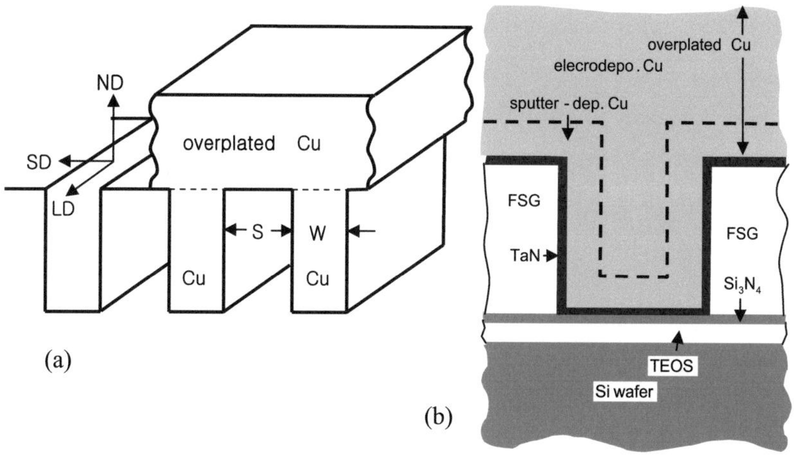

Figure 18. (a) Structure of damascene trenches, where ND, SD, and LD stand for normal direction, sidewall direction, and line direction of trench, respectively. S, W, and Cu indicate space, trench width, and Cu electrodeposit, respectively. (b) Schematic of cross section of electroplated Cu over damascene trench. FSG and TEOS mean fluorinated silicate glass and tetraethylorthosilicate, respectively [41].

Figure 19 shows plan-view electron backscatter diffraction (EBSD) maps of the surface of Cu electrodeposits in 500 nm/500 nm width/space patterned trenches obtained at 2.5 h, 3.5 h, 5.5 h, and 12.5 h at room temperature after Cu electrodeposition. The black area in the EBSD maps indicates the low pattern-quality region. In the low pattern-quality region, the Kikuchi patterns cannot be obtained because the region is too distorted or grains there are too small [41]. The spatial resolution of the EBSD system in the experiment was about 10 nm. Therefore, the grain size in the black area is estimated to be not larger than 15 nm [41]. It took about 30 min to scan from the top line to the bottom line of one specimen to obtain its EBSD map [41]. For the 2.5 h specimen in Figure 19, EBSD scanning started 2.5 h after Cu electrodeposition and finished 3 h after Cu electrodeposition. It can be seen that the black area decreases or the measurable EBSD area increases with increasing time [41]. For the 2.5 h specimen, the measurable EBSD area is seen to increase, even while it is scanned for 30 min. The black vertical lines in the EBSD maps in Figure 19 are related to the blanket area in the trench pattern [41]. The black area including the black line width decreases with increasing time. These results indicate that the grain growth rate is inhomogeneous. The grain growth rate is faster in the pattern area than in the blanket area [41]. According to Lingk and Gross [42], the grain growth initiates at the upper corners of the trench plugs. The number density of corners is higher in the pattern area than in the blanket area. Therefore, the grain growth is faster in the pattern area than in blanket area [41]. The black lines which are related to blanket area are not clearly seen in Figure 19 because of the smaller width of blanket area [41]. To conclude, the initial grain size is not larger than 15 nm and the grain growth occurs even at room temperature. This phenomenon is related to the texture evolution of interconnects, as will be discussed later [41].

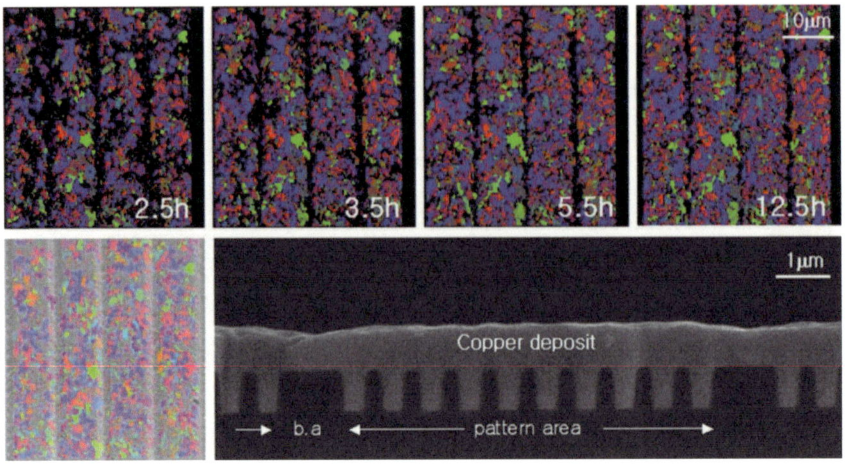

Figure 19. Plan-view EBSD maps of surface of Cu-plated 500 nm/500 nm width/space patterned trenches as a function of time at room temperature following electrodeposition. Black area in EBSD maps is low pattern-quality region. (Left below EBSD maps) Plane-view EBSD map superimposed on corresponding SEM micrograph. (Right below EBSD maps) SEM micrograph of plane normal to trench lines, in which b.a stands for blanket area. Green and blue (light and dark in black-and-white figure) areas in EBSD maps indicate <110>//ND and <111>//ND oriented regions, respectively [41].

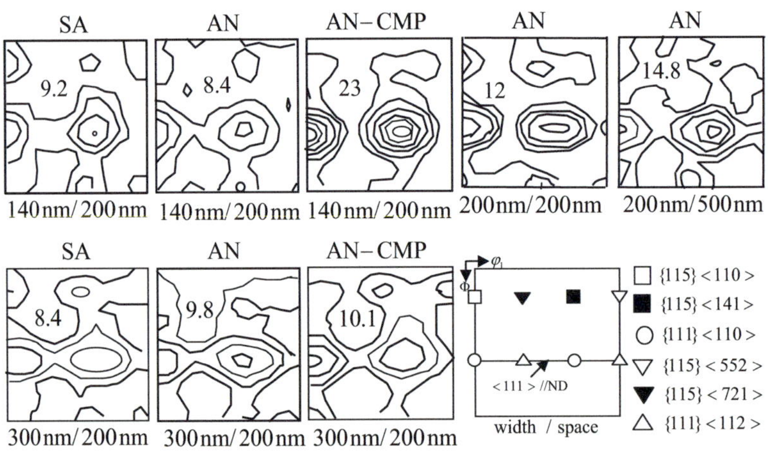

Figure 20. Plan-view ODFs ($\varphi_2 = 45°$) of Cu electrodeposits over various trench/space gratings. SA means self-annealed deposits with over-layer. AN means deposits with over-layer after annealing for 10 min at 200°C. AN-CMP means deposits after annealing for 10 min at 200°C and removing over-layer by CMP. Numerical values in ODFs indicate the maximum orientation density [41].

Figure 20 shows the plan-view ODFs of the surfaces of Cu electrodeposits on various widths/spaces trenches and blanket wafer after self-annealing at room temperature and after annealing for 10 min at 200°C [41]. The textures of all the specimens except the blanket specimen are similar and can be approximated by the {111}<110> texture as a major component and its twin components {115}<110> and {115}<141> as minor ones [41]. Here {hkl}<uvw> means that {hkl} is the surface plane (<hkl>//ND) and <uvw> is the line direction (LD). The texture of the deposit on blanket wafer is approximated by a <111>//ND texture [41]. Figure 21 shows more annealing texture data on different trench width/space gratings. The self-annealing textures of the 200 nm/200 nm and 240 nm/200 nm width/space patterned trenches can be approximated by a major {111}<110> component and a minor {115}<110> and {115}<141> components, which are twin components of {111}<110>, whereas the texture of the 1000 nm/200 nm width/space specimen is approximated by a major {111}<112> component and a minor {115}<552> component, which is a twin component of {111}<112>. The textures of the 500 nm/200 nm, 2000 nm/200 nm, 500 nm/500 nm, and 500 nm/1000 nm specimens are seen to be mixtures of {111}<110> and {111}<112>, contributions of which vary with specimen.

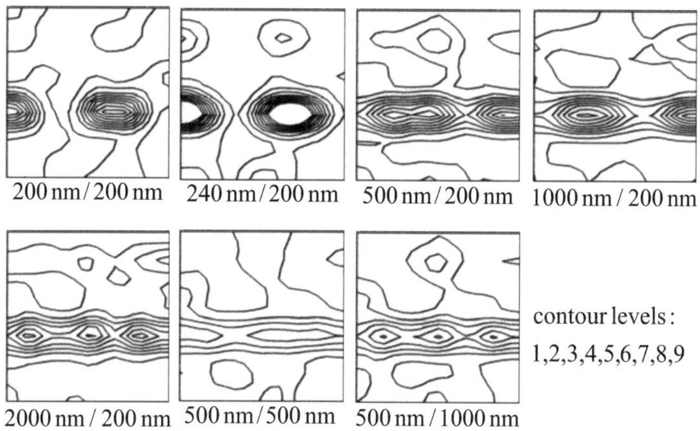

Figure 21. ODF ($\varphi_2 = 45°$) of self-annealed Cu electrodeposits over various trench width/space gratings [43].

Figure 22 shows annealing texture data of Cu damascene interconnects with 0.2-μm space [44]. The textures of the specimens of 0.2 and 0.24 μm in trench width can be approximated by {111}<110> as the major component and its twin components {115}<110> and {115}<141> as minor ones [44]. The textures of the specimens of 0.5 and 1 μm in width can be approximated by {111}<112> with weak <111>//ND [44]. As the width increases, the texture is increasingly of fiber character, with the orientation density decreasing [44]. It is noted that the textures are almost independent of annealing conditions. The {111}<112> texture prevails among the CMP specimens. The 4- and 6-μm-wide specimens show textures approximated by mixtures of diffused {111}<110> and {111}<112> [44]. It is interesting to note that the textures measured over the over-plated layer are similar to those of the CMP-specimens [41]. In order to understand the phenomenon, the structures of the cross sections normal to the trench lines were measured by EBSD [41]. The EBSD maps of the cross sections normal to the line direction (LD) of various trench specimens (Figure 23) show that most grains along with their twins stretch from the bottom of trenches to the surface of over-plated layer. One trench is hardly filled by multiple grains in the normal direction (ND) of trench lines. This is the reason why the plan-view orientations of the copper deposits are almost the same before and after CMP [41].

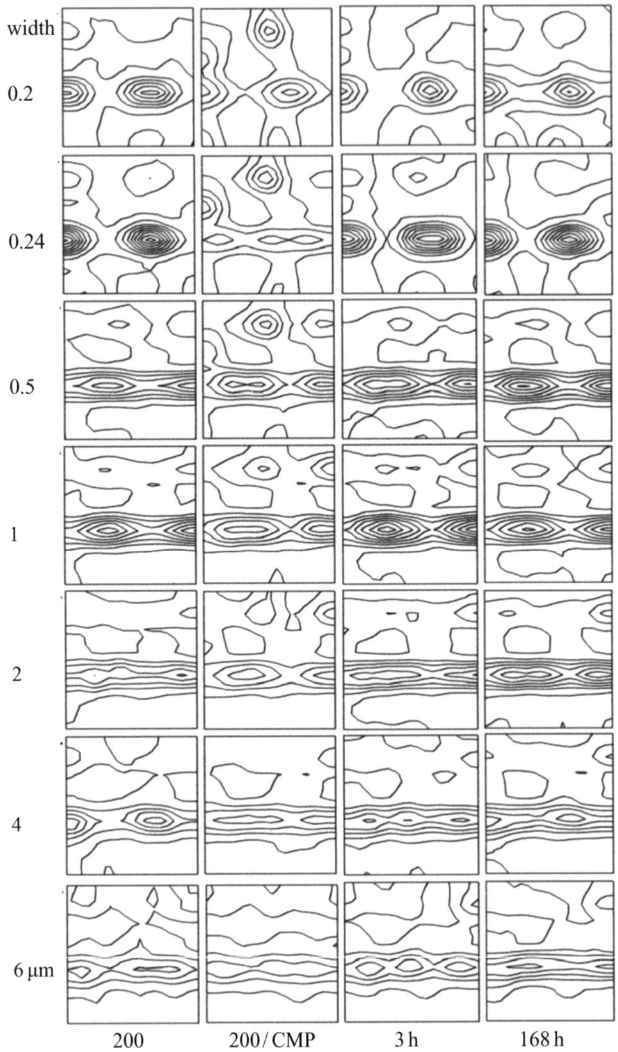

Figure 22. ODF ($\varphi_2 = 45°$) of Cu electrodeposits over various trench gratings with 0.2-μm space. Trench width ranges from 0.2 to 6 μm. Specimen (200) annealed for 10 min at 200°C following deposition, (200/CMP) specimen (200) after CMP, (3h) self-annealed for 3 h, (168 h) self-annealed for 168 h [44].

In order to understand the texture results, we must first understand the evolution of the {111}<110> and {111}<112> texture components. The evolution of {111}<110> was first explained by Lee and Lee [43]. Cu deposits electroplated on Cu seed layers have the <111> texture influenced by a strong <111>//ND texture of the seed layers. The seed layers are deposited on the TaN barrier material by sputtering and have a strong <111>//GD texture (GD: the growth direction) [45]. In order to improve filling of Cu deposit in trenches, organic additives are added to the electroplating bath, which hinder growth of electrocrystallites, resulting in fine grain sizes. The initial grain size in the electrodeposits is estimated to be less than 15 nm, resulting in little dislocation density in the deposit [43]. Therefore, their annealing texture has nothing to do with dislocations. In this case, the annealing texture is controlled by interface energy, grain boundary energy and mobility, surface energy, and strain energy [46,47], and the existing grains that have lower interface, grain boundary, surface, and strain energies and higher grain boundary mobility are in a better position to grow and dominate the annealing texture. This growth texture does not need nucleation for the texture evolution [44]. When a polycrystalline deposit having a fiber texture is annealed under balanced stresses, the annealing texture is likely to of fiber character, whether the annealing texture differs from the deposition texture or not [44]. When the deposit is thick enough to neglect the surface energy and/or the interface energy between the deposit and substrate compared with the internal energy due to stresses, whatever sources they may originate from, the annealing texture will be dominated by the stresses [44]. When the stresses are caused by dislocations, the evolution of annealing texture (recrystallization texture) is well explained by the SERM theory. In the SERM theory [1], the recrystallization texture is determined such that the absolute, maximum stress direction (AMSD) due to dislocation array aligns, at least, with the minimum Young's modulus direction (MYMD) in recrystallized grains, whereby the strain energy release can be maximized. According to the SERM theory, grains whose MYMD is closest to AMSD grow in preference to others. When a polycrystalline material is strained, the strain of each grain can be approximated by the strain of the material, but its strain energy varies because its elastic constants vary with the crystal direction [44]. For simplicity of explanation, we consider the uniaxial tension of a bicrystal consisting of the crystals A and B as shown in Figure 24. Young's moduli E_A and E_B of the crystals A and B differ because of their different orientations. The strain energy densities, w_A and w_B, of the crystals A and B are given by

$$w_A = E_A \varepsilon^2 / 2 \text{ and } w_B = E_B \varepsilon^2 / 2 \tag{25}$$

where ε is the tensile strain. For $E_A > E_B$, the strain energy of A is larger than that of B. A polycrystalline film shown in Figure 25 is considered. Grains whose MYMD is aligned with AMSD have the minimum strain energy. Therefore, the growth of these grains will be preferred. When these grains grow, the strain-energy-release of the film becomes maximized, and the annealing texture is determined by grains whose MYMD is parallel to AMSD. The grain growth rate in a high strain energy region is expected to be higher than in a low strain energy region because the grain growth rate in the high strain energy region reduces the strain energy more rapidly [44].

The strain energy distributions and AMSD in the structures with damascene trenches filled with copper (Figure 18) are considered. The grain size of the copper deposits is of the order of 10 nm as already mentioned [44]. The nanocrystalline deposits have highly strained grains, possibly under tensile stresses because grain boundaries occupy a substantial portion of the volume and are of open structure [44]. Nanocrystalline copper electrodeposits have tensile residual stresses [48-51]. To simulate qualitative distributions of the tensile internal stress field and strain energy in the trenches and their surroundings, Lee et al. [44] have performed a numerical calculation using a finite element code, ABAQUS 6.2. Isoparametric quadratic

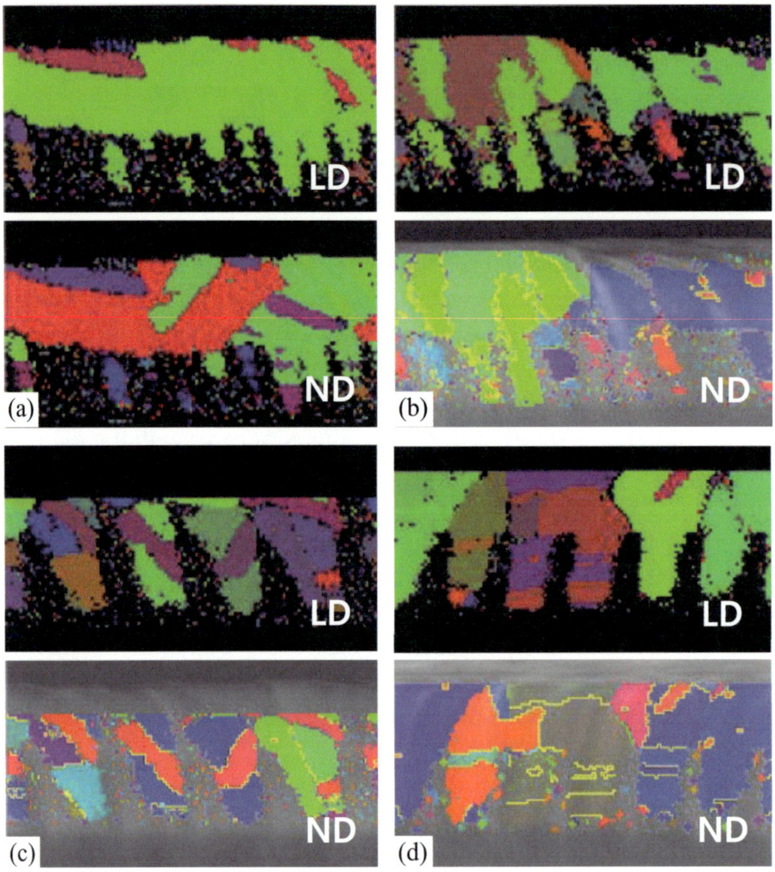

Figure 23. EBSD maps of vertical trench sections. (a) 140 nm/140 nm trench/space grating after self-annealing, (b) 200 nm/200 nm trench/space after self-annealing, (c) 500 nm/500 nm trench/space after self-annealing, (d) 500 nm/500 nm trench/space after annealing for 10 min at 200. Upper and lower figures indicate orientations along line direction and normal direction, respectively. Yellow boundaries indicate twin boundaries [41].

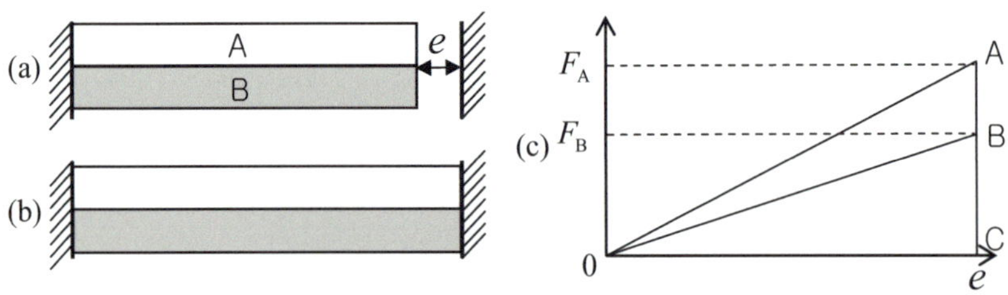

Figure 24. Bicrystal is deformed by displacement e (a) and its both ends are fixed (b). Strain energy of crystals A and B are represented by OAC and OBC (c). Crystals A and B are loaded by forces F_A and F_B, respectively [44].

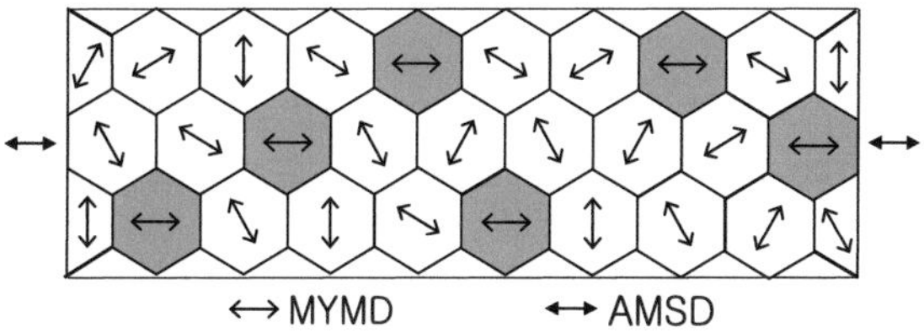

Figure 25. Dark grains whose minimum MYMD is parallel to AMSD grow preferably during annealing and their orientation determines annealing texture [44].

elements were used under plane strain condition, no strain along LD, and the material was assumed to be an isotropic thermo-elastic solid. The cross-sectional structure of a damascene copper deposit including its surrounding materials is shown in the upper right of Figure 26 [44]. The internal stress and strain energy distributions were calculated from the thermal expansion mismatch due to cooling by 40°C. Physical properties of Cu and its surrounding materials are given in Table 1. The strain-energy densities and the maximum principal-stress distributions and directions in the copper deposited in and over three different damascene trenches (a =100 nm and b =200 nm; a =250 nm and b =350 nm; a =400 nm and b = 500 nm) were calculated. One of them is shown in Figure 26. Most of the maximum principal-stress directions in trenches were parallel to the LD of trenches except the high strain-energy regions close to their upper corners [44]. The largest principal-stress direction is in a form of an in-plane stress at the upper corners and is inclined to the deposit surface because of the notch effect of the upper corners [44]. The inclination angle is less than 45° because the thickness of the deposit over the trenches is shorter than the dimension along the surface and less constrained along the ND [44]. It is also noted that the strain energy densities are highest at the upper corners of the trenches. Therefore, the grain growth rate in this region is higher than other regions [44]. This is in agreement with the Lingk and Gross's experimental observation [42] that the grain growth initiates at the upper corners of the trenches [44].

It was observed that the volume fraction of the region, in which the maximum principal-stress directions are parallel to the LD, increases and the strain energy densities at the upper corners increase with increasing width of trenches. The sense of stresses changed, while the strain energy density distributions remain unchanged when the specimens are heated. When the grain growth rate in the high strain-energy region is dominant, the annealing texture is determined such that AMSD at the upper corners of trench is parallel to MYMD of Cu. For the <111>//ND Cu deposit, the MYMD of Cu, the <100> directions, are not on the {111} planes. The directions that are on the {111} planes and at the minimum angle with the <100> directions are the <112> directions as shown in Figure 27. Therefore, grains with the <112> directions have the minimum strain energy among the <111>//ND grains and grow preferably. For Cu deposited in trenches (Figure 18), the <112> directions are parallel to SD. In this case, the <110> directions are parallel to the LD and the {111}<110> oriented grains grow preferably during annealing, resulting in the {111}<110> annealing texture [44]. Similarly, the grain growth rate in the trench, where the AMSD is parallel to LD, is dominant, and the {111}<112> annealing texture is obtained [44].

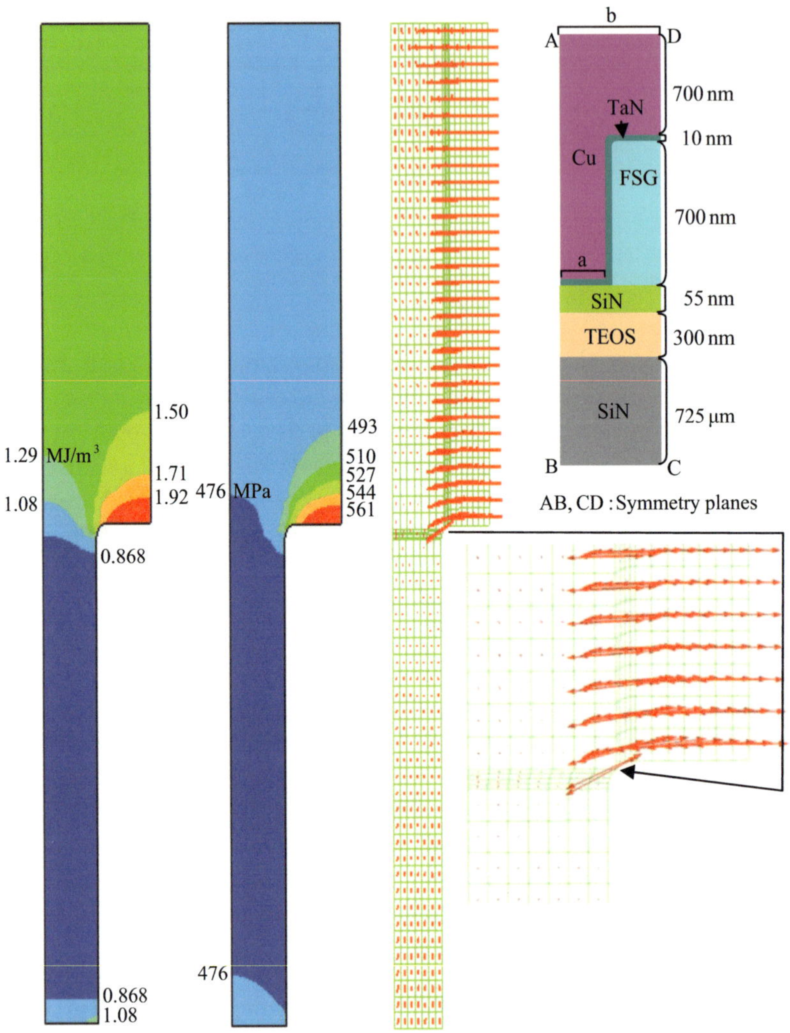

Figure 26. Strain energy density and maximum stress distributions and maximum stress directions (from left) in Cu deposited in and over damascene trench with a = 100 nm and b = 200 nm in Figure 18 [44].

Material	Si(100)	Cu [54]	TEOS [55]	Ta [56]	SiN [57]	FSG [58]
CTE, ppm/K	2.61 [52]	17.7	1.0	6.5	3.2	0.94
Young's modulus, GPa	131.0 [53]	104.2	59.0	185.7	220.8	71.7
Poisson's ratio	0.278 [53]	0.352	0.16	0.34	0.27	0.16

Table 3. Physical properties of Si(100), Cu, TEOS, Ta, SiN, and FSG.

For the 0.2- and 0.24-μm-wide specimens, the {111}<110> annealing texture was obtained, possibly because the grain growth rate in the high strain-energy region is dominant [44]. As

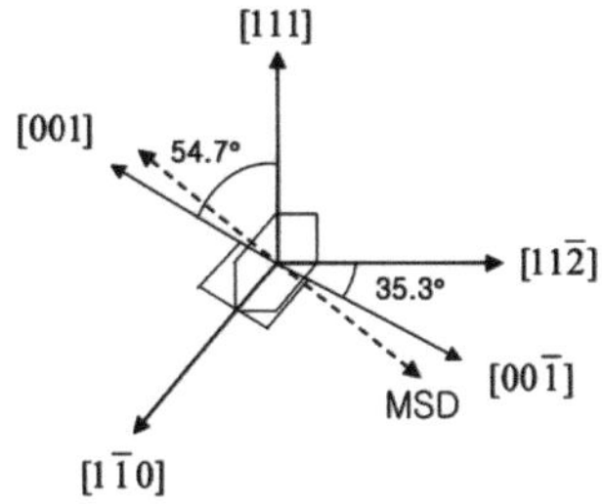

Figure 27. Relationship among [111], [1-10], [11-2], and [001] directions. Maximum stress direction (MSD) and [111], [001], and [11-2] directions are on same plane [2].

the width of trench increases, the grain growth in the trench, where the MSD is parallel to the LD, increasingly prevails, although the strain energy density is low, resulting in the {111}<112> texture [44]. This is supported by the fact that the {111}<112> texture prevails among the CMP specimens (200/CMP in Figure 22). It is noted that when the {111}<112> texture forms, the fiber component is always found, implying that the {111}<110> component cannot be excluded [44]. Up to 2 μm in width, the textures are similar regardless of annealing conditions (10 min at 200°C and 3 h and 168 h at room temperature). The grain growth in the copper deposits is completed in a few hours, resulting in similar textures in the deposits after annealing for 3 h and 168 h at room temperature [44]. Heating at 200°C could develop the sense of stresses opposite to that at room temperature. However, the above discussion still holds. Therefore, the textures of Cu deposits are similar regardless of annealing conditions. As the trench width further increases to 4 and 6 μm, the textures are characterized by mixtures of the {111}<110> and {111}<112> textures or near fiber textures [44]. The relatively high density near φ=15° and φ_1=45° of the CMP specimens seems to be related to the texture of trench walls [44]. This needs further studies.

5. Annealing textures of nanocrystalline copper thin films

Lee et al. [59] studied texture analysis of abnormally grown (abnormal) grains in nanocrystalline Cu films using an automatic orientation- and phase-mapping technique for transmission electron microscopy (TEM). Cu films were deposited by DC sputtering at room temperature onto amorphous SiNx/Si TEM grids with a viewable area of 100x100 μm^2 (Figure 28). They traced the texture evolution of abnormal grains with post-deposition annealing temperature, for which textures of aggregates of grains smaller and larger than specific grain sizes in four types of specimens (as-deposited, annealed for 150 min at 125°C, annealed for 60 min at 300°C, annealed for 60 min at 500°C) were examined. The textures of the aggregate of grains smaller than and larger than a reference grain size are designated as matrix texture and abnormal grain growth (AGG) texture, respectively. The texture of the whole grains, not size-selected grains, is termed overall texture. For the as-deposited film, the overall and matrix textures were dominated by <110>//ND, but <100> //ND grains tended to grow abnormally.

After annealing at high temperatures from 125 to 500°C, the overall and matrix textures were changed to major <112>//ND. The AGG texture was dominated by <112>//ND at 125°C and 300°C. On the other hand, at 500°C, the major AGG texture component was <111>//ND.

In the as-deposited film, <100>//ND grains and a smaller fraction of <111> //ND grains underwent abnormal growth, which can be explained by elastic strain energy minimization and surface energy minimization, respectively. After annealing at 125 and 300°C, AGG developed a strong <112>//ND texture component. After annealing at 500°C, abnormal grains tended to have a major <111>//ND texture component, which is simply explained by surface energy minimization. As annealing temperature increases above 125°C, a strain due to differential thermal expansion would build up, and some grains would undergo plastic deformation before other grains, which is suggested to depend on their Taylor factors. Grains with lower Taylor factors will deform before those with higher factors. Plastically deformed grains are expected to be consumed by grains still remaining in the elastic regime due to the stored strain energy caused by plastic deformation. Among the grains in the elastic regime, grains with the minimum biaxial elastic modulus are likely to grow at the expense of other grains. The hypothesis successfully explained the texture evolution observed after annealing at 125 and 300°C. The attempt also seemed to explain the <112>//ND texture component developed in the Al alloy thin films [69] with little ambiguity.

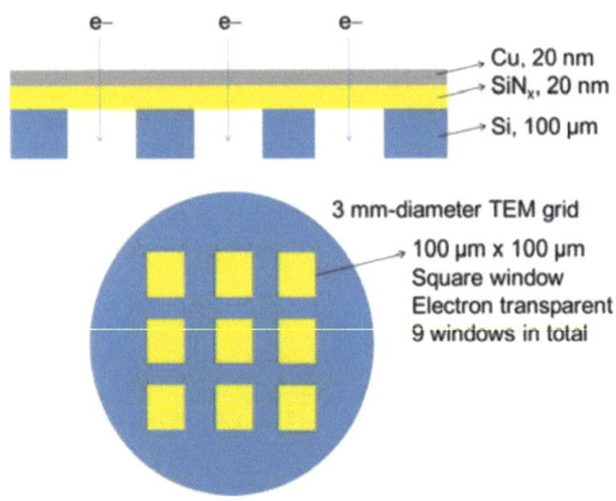

Figure 28. Schematic diagram of Cu films deposited on a grid for in situ TEM (back-etched Si wafer with an amorphous SiN_x membrane [59].

The Cu target purity was 99.999 %. The thickness of the SiN_x membrane which was deposited on 100-μm-thick Si was 20 nm. During deposition, the chamber pressure and the process gas pressure were 2×10^{-7} Torr and 8 mTorr of Ar, respectively. The sputtering power was 2500 W, exhibiting a sputtering rate of ~2 nm/s. The Cu film thickness was determined to be 20 nm. The specimens were subsequently annealed for up to 150 min isothermally at 125°C, and for up to 60 min at 300°C and 500°C in JEM 3011 operating at 300 keV (0.2-nm point-to-point resolution). The heating and cooling rates used were ~5°C/s. The as-deposited film and deposits annealed for 150 min at 125°C, for 60 min at 300°C, and for 60 min at 500°C were examined for the analysis of texture evolution with post-deposition annealing temperature by using an orientation and phase mapping technique (ASTAR) attached to a field-emission TEM (JEM 2100F) [59]. This technique produced orientation maps similar to the EBSD map obtained in a scanning electron microscope, but exhibits a higher spatial resolution and reasonable acquisition speed for TEM [59]. The setup of the ASTAR system was adjusted to have 2.4 nm in spot size of the electron beam and 4.1 nm in scanning step size.

The raw data obtained from the system was processed by using Orientation Imaging Microscopy™ (OIM™) 6.2 software (EDAX, Draper, U.S.A.) to demonstrate textures of the Cu films in terms of area fraction. The film texture is represented by <*hkl*>//ND, which indicates the <*hkl*> direction of the deposit is parallel to the deposit-surface normal direction (ND). For the matrix and AGG textures, three times the mean grain size was applied as the reference grain size for the as-deposited film and the film annealed at 125°C. As annealing temperature increased further, as will be shown later, the mean grain size did not increase much but larger grains grew fast, leading to the broadened grain size distributions at 300°C and 500°C. Thus a reference of five times the mean grain size was added for the films annealed at these two temperatures.

Orientation data were processed by grain unit. Grains were identified by 5°-misorientation criterion and grains under two pixels were replaced by surrounding pixels to minimize the misindexing influence. For grain size analysis, grains meeting at twin boundaries were considered as one grain [59]. The strength of a texture component [f(g) value] in ODF space cannot reflect the volume fraction of the component directly because there exist overlapped equivalent points of the component in ODF space, which results in the concept of multiplicity. The higher the number of overlapped equivalent points of a texture component in ODF space, the larger its f(g) value [59]. For the quantitative texture analysis of grain growth, they directly calculated area fractions of each major texture component from ASTAR orientation data. Ten-degree criterion is introduced for the determination of each data point as the major texture components because over this criterion, areas for the <112>//ND and <111>//ND components are overlapped.

Figure 29 a shows orientation mapping image, the corresponding grain size distribution and calculated result of area fraction of the as-deposited film. The area fractions correspond to partition fraction in the data captions. Total fraction means a fraction of each component with reference to the whole grains, not size-selected grains. Though not intentionally heated, the film already revealed AGG behavior, as indicated by an arrow in the grain size distribution. For overall deposition texture, the <110>//ND grains occupied the largest area fraction. To extract the AGG texture from overall texture, area fractions of the major components for aggregates of grains smaller than and larger than three times the mean grain size (~35 nm) were calculated (Figure 29 b). The texture of grains smaller than the reference size (matrix texture) shows a major <110>//ND component (Figure 29 b, left). On the other hand, for the AGG texture (Figure 29 b, right), the area fraction of the <110>//ND grains decreases, and those of <100>//ND, and <111>//ND grains greatly increased as compared with the matrix texture, As a result, <100>//ND, <111>//ND, and <110>//ND grains occupied similar fractions, as represented in the bar graph (Figure 29 b).

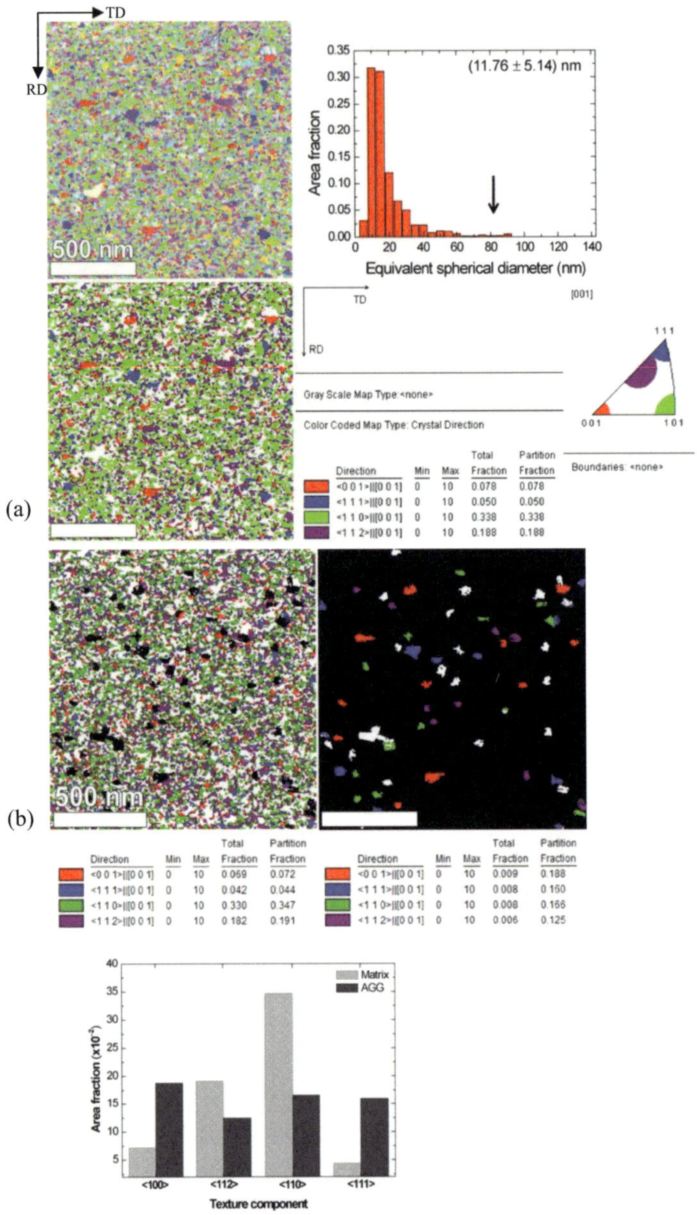

Figure 29. (a) Orientation mapping image and grain size distribution of as-deposited film (upper panel). Average grains size (in diameter) and standard deviation are specified. Area fraction map for the overall data is shown in the bottom left with its color code (bottom right). (b) Area fraction maps of aggregates of grains smaller (left) and larger (right) than reference size of 35.28 nm (3 times mean grain size). Area fractions of texture components in matrix and AGG textures are plotted in bar graph (bottom) [59].

After annealing at 125°C, the overall texture (Figure 30 a, right) was observed to be mainly covered by <112>//ND grains, whereas the area fraction of <110>//ND grains was strongly decreased from ~0.34 for the as-deposited film to ~0.18. Such tendencies were also reflected on the matrix texture (Figure 30 b, left). Also for the AGG texture, <112>//ND grains occupied the largest area fraction, exhibiting no noticeable difference in area fraction between the matrix and AGG textures. <100>//ND and <111>//ND grains even increased their area fractions in the AGG texture (Figure 30 b, bar graph). For the deposit annealed at 300°C, the overall, matrix and AGG textures were observed to be mainly <112>//ND (Figure 31). However, at variance with the case of the film annealed at 125°C, after annealing at 300°C, the area fraction of <112> grains drastically increased in the AGG texture for both references of three times (Figure 31 b, bar graph) and five times (Figure 31 c, bar graph).

According to a model for development of orientation of vapor deposits [60,61], under the condition of a relatively high vapor concentration adjacent to the deposit induced by high substrate temperatures and evaporation rates, the texture of vapor deposits is dominated by the orientation that places higher surface energy crystal planes (highly stepped surface) normal to the growth direction because, in this regime, grains with highly-stepped surfaces can be in a better position to grow than those with lower-energy surfaces (lowly-stepped surfaces with broad terraces) normal to the growth direction. In contrast, under the condition of a lower vapor concentration adjacent to the deposit induced by relatively low substrate temperatures and evaporation rates, the film texture will be dominated by grains with lower-energy surface normal to the growth direction because the terrace area to be covered by atoms is smaller per unit apparent area. The <110> deposition texture in this study implies that the deposition condition belongs to the regime of a relatively high vapor concentration adjacent to the deposit.

The annealing texture of thin films is determined by energies due to the surface, the film/substrate interface, the extrinsic stress (plane stress-equibiaxial strain or inhomogeneous stress distribution), and intrinsic stress due to dislocations [47]. In this system, the interface energy is unlikely to control the annealing texture because the SiN_x film between the Cu deposit and the Si substrate is amorphous. The initial average grain diameter of the copper deposit was smaller than 20 nm (Figure 29 a). Therefore, the Cu grains are likely to have little dislocation density under no stress. Since the grain size and thickness of the deposit are similar, the surface energy could be important. However, surface energy minimization predicts the <111>//ND texture, and thus cannot explain the <100> and <112> AGG textures (Figures 29-32 and Table 4).

When the layered specimen is heated to 125, 300, and 500°C, the Cu/SiN_x layer can be subjected to a thermal strain, due to the difference in thermal expansion coefficient between the Cu/SiN_x layer and the Si substrate. In this case the Cu/SiN_x layer should be in a plane stress, equibiaxial strain state because the 20-nm-thick Cu/20-nm-thick SiN_x layer is very thin, as compared with the 100-μm-thick Si substrate, and thus its strain is likely controlled by the Si substrate whose thermal expansion is isotropic.

As for the deposit annealed at 125 and 300°C, the deposit annealed at 500°C revealed that the overall and matrix textures were dominated by <112>//ND (Figure 32). For the AGG texture, in the case of a reference of three times (Figure 32 b, bar graph), <112>//ND grains occupied the largest area fraction with that of <111>//ND occupying the second largest area fraction. For a reference of five times, the situation was reversed with <111>//ND grains making up the

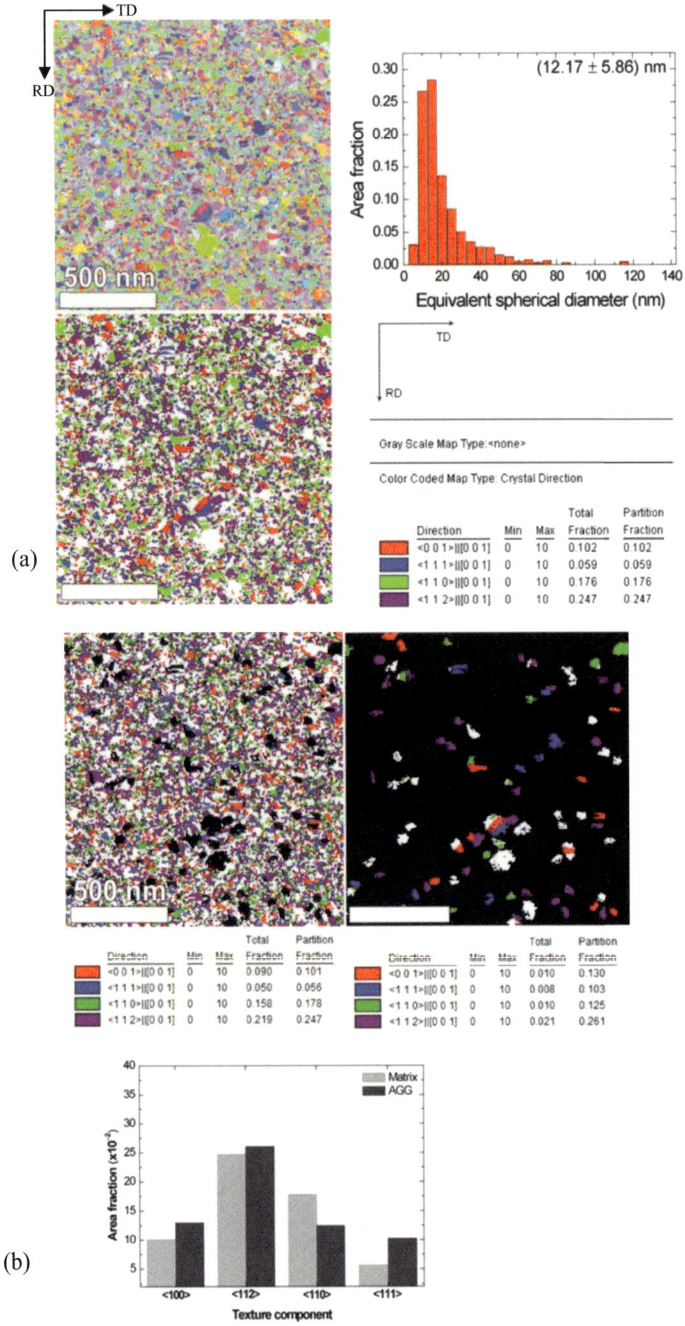

Figure 30. (a) Orientation mapping image, grain size distribution (center), and area fraction map (right) of the deposit annealed at 125°C. (b) Area fraction maps of the aggregates of grains smaller (left) and larger (right) than a reference size of 36.51 nm (3 times the mean grain size) [59].

Temperature	Overall	Matrix	AGG
RT	110	110	100
125 °C	112	112	112
300 °C	112	112	112
500 °C	112	112	111

Table 4. Summary of overall, matrix, and AGG textures for all temperatures examined [59].

largest fraction followed by <112>//ND grains. The summary of overall, matrix and AGG textures for all the temperatures examined is given in Table 4.

Since, for fcc metals, the surface energy increases in order of {111}, {100}, {112}, and {110} (Figure 16), the main <110>//ND component in the overall and matrix deposition textures (Figure 29, Table 4) is not explained by surface energy minimization, but should be considered by growth kinetics. The thermal strain, ε_{th}, due to differential thermal expansion is calculated by

$$\varepsilon_{th} = \int_{T_0}^{T} \left(\alpha_s - \alpha_f\right) dT \tag{26}$$

where T and α are temperature and linear thermal expansion coefficient, respectively; and the subscripts 0, s, and f indicate initial state, substrate, and film, respectively. The linear thermal expansion coefficients of Cu, amorphous Si_3N_4 for the SiN_x layer, and Si are 16.5×10^{-6} K^{-1} [http://www.webelements.com], 4.5×10^{-6} K^{-1} [62], and 2.6×10^{-6} K^{-1} [http://www.webelements.com], respectively. During annealing, the Cu layer is likely to undergo an equibiaxial compressive strain of 1.39×10^{-3} [= (16.5–2.6)(125–25) × 10^{-6}] at 125°C, 3.82×10^{-3} [= (16.5–2.6)(300–25) × 10^{-6}] at 300°C, and 6.60×10^{-3} [= (16.5–2.6)(500–25) × 10^{-6}] at 500°C. The strain energy density (the strain energy per unit volume) w of a film under the plane stress, equibiaxial strain (ε) state is given by [35,36]

$$w = \varepsilon^2 M_\varepsilon \tag{27}$$

where M_ε is the biaxial elastic modulus of the film under the plane stress, equibiaxial strain state. The modulus M_ε for a cubic-symmetry material is expressed as Eq. 28 [35,36],

$$M_\varepsilon = \frac{1}{2} \cdot \frac{S'_{11} + S'_{22} - 2S'_{12}}{S'_{11}S'_{22} - \left(S'_{12}\right)^2} \tag{28}$$

where the compliances S'_{11}, S'_{12}, and S'_{22} are given by Eqs. 29, 30, and 31, respectively [35,36].

$$S'_{11} = S_{11} + \left[S_{44} - 2(S_{11} - S_{12})\right]\left(a_{11}^2 a_{12}^2 + a_{12}^2 a_{13}^2 + a_{13}^2 a_{11}^2\right) \tag{29}$$

$$S'_{12} = S_{12} - (1/2)\left[S_{44} - 2(S_{11} - S_{12})\right]\left(a_{11}^2 a_{21}^2 + a_{12}^2 a_{22}^2 + a_{13}^2 a_{23}^2\right) \quad (30)$$

$$S'_{22} = S_{11} + \left[S_{44} - 2(S_{11} - S_{12})\right]\left(a_{22}^2 a_{21}^2 + a_{21}^2 a_{23}^2 + a_{23}^2 a_{22}^2\right) \quad (31)$$

For the [*hkl*]//ND crystals, the transformation matrix a_{ij} is expressed as Eq. 32 [35,36].

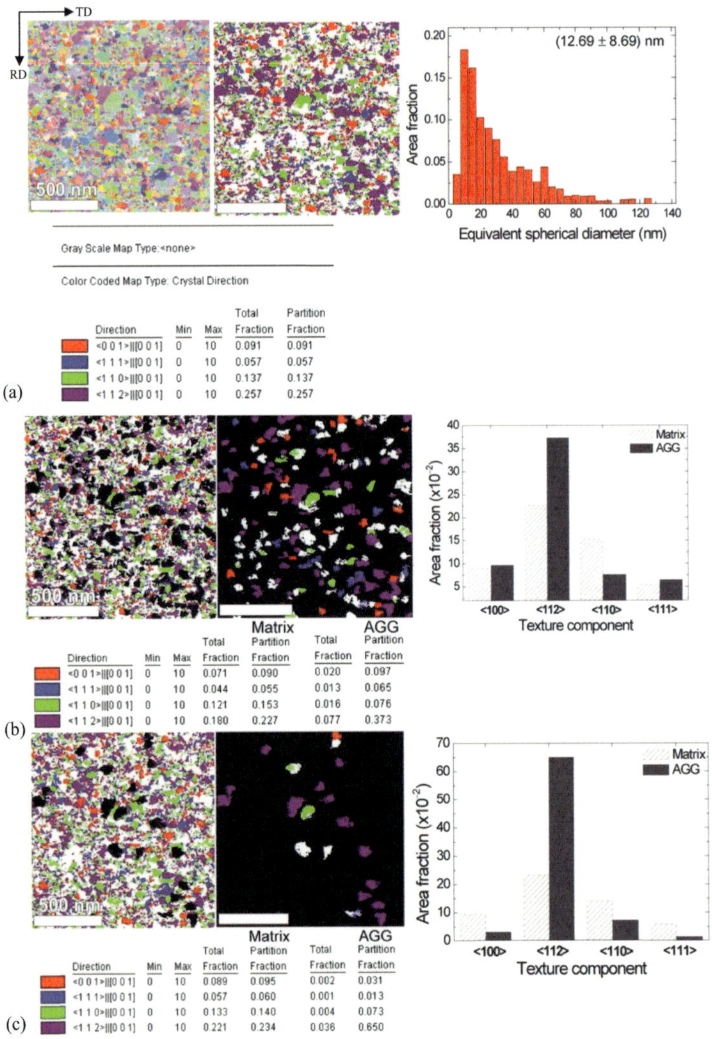

Figure 31. (a) Orientation mapping image, grain size distribution, and area fraction map of deposit annealed at 300°C. (b) Area fraction maps of the aggregates of grains smaller (left) and larger (right) than reference size of 38.07 nm (3 times mean grain size). (c) Area fraction maps of the aggregates of grains smaller (left) and larger (right) than reference size of 63.45 nm (5 times mean grain size) [59].

http://dx.doi.org/10.5772/59724

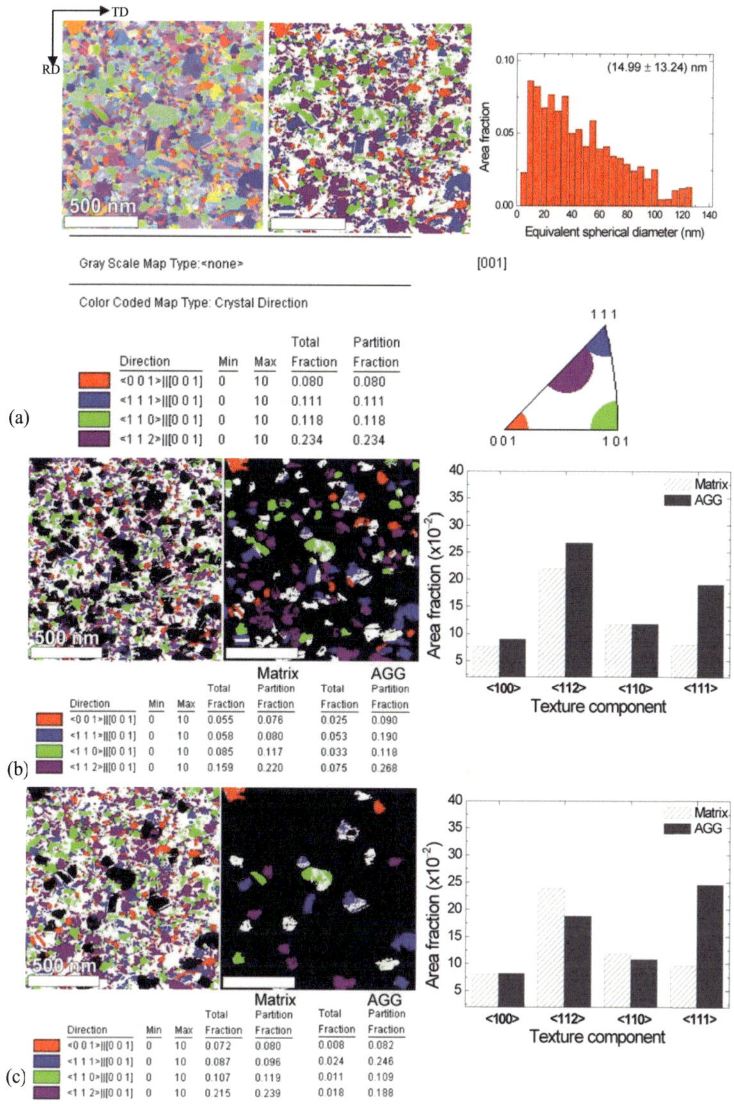

Figure 32. (a) Orientation mapping image (left), grain size distribution (center), and area fraction map (right) of the deposit annealed at 500°C. (b) Area fraction maps of aggregates of grains smaller (left) and larger (right) than reference size of 44.97 nm (3 times mean grain size). (c) Area fraction maps of aggregates of grains smaller (left) and larger (right) than reference of 74.95 nm (5 times mean grain size) [59].

For copper, S_{11} = 0.015635, S_{44} = 0.01368, S_{12} = –006587 GPa^{-1} at 400 K and S_{11} = 0.016718, S_{44} = 0.014472, S_{12} = –0.00691 GPa^{-1} at 550 K [63]. It follows from these data that M_ε of the <100>//ND grain (M_ε <100>) = 110.5 GPa, M_ε <112> = 200 GPa, M_ε <110> = 227 GPa, M_ε <111> = 255.1 GPa at 400 K, and M_ε <100> = 102 GPa, M_ε <112> = 182 Ga, M_ε <110> = 206.2 GPa, M_ε <111> = 230.2 GPa at 550 K. In fact, M_ε <100> and M_ε <111> are the minimum and maximum values, respectively. Note that, as temperature increases, strain energy anisotropy also increases.

$$a_{ij} = \begin{bmatrix} \frac{-k}{\sqrt{h^2+k^2}} & \frac{h}{\sqrt{h^2+k^2}} & 0 \\ \frac{-hl}{\sqrt{h^2+k^2}\sqrt{h^2+k^2+l^2}} & \frac{-kl}{\sqrt{h^2+k^2}\sqrt{h^2+k^2+l^2}} & \frac{h^2+k^2}{\sqrt{h^2+k^2}\sqrt{h^2+k^2+l^2}} \\ \frac{h}{\sqrt{h^2+k^2+l^2}} & \frac{k}{\sqrt{h^2+k^2+l^2}} & \frac{l}{\sqrt{h^2+k^2+l^2}} \end{bmatrix} \quad (32)$$

As revealed in Figure 29, the as-deposited film already underwent AGG with the <100> and <111> texture components contributing alike. It may imply that during deposition the substrate temperature increased to temperatures above room temperature, thus suggesting that the AGG texture is affected by thermal strain build-up and thus controlled by elastic strain energy minimization. In fact it is well documented that, though the substrate is not intentionally heated, it is warmed by the deposition itself [64]. Considering that <100> grains have the minimum biaxial elastic modulus, the evolution of the AGG texture of <100> for the as-deposited film (Figure 29 b and Table 4) is explained by elastic strain energy minimization. The <111> component in the AGG texture is attributed to surface energy minimization effect, indicating that in the as-deposited film, two representative driving forces (surface energy minimization and strain energy minimization) were competing.

One of the questions to be addressed is why the overall, matrix, and AGG textures after annealing at 125 and 300°C were dominated by the <112> component (Figures 30 to 32 and Table 4). The component seems not related to elastic strain energy minimization because <100> grains have the minimum biaxial elastic modulus. However, as will be discussed below, if we assume that the film system is elastic–plastic and introduce the concept of the Taylor factor, we will be forced to know that the evolution of the <112> component is understood in terms of elastic strain energy minimization. During annealing at 125 and 300°C, the deposit is calculated to undergo the equibiaxial compressive strains of 1.39×10^{-3} at 125°C and $3.82.60\times10^{-3}$ at 300°C, as noted before, and at such high temperatures the critical resolved shear stress on active slip systems is expected to be lowered. Therefore, the deposit may undergo plastic deformation. If every grain undergoes the same strain as the deposit does, operating slip systems and shear strains on the slip systems can differ depending on the orientation of each grain. In order to accomplish an increment of plastic strain, $d\varepsilon_{ij}$, the shear strains, $\sum_k d\gamma^{(k)}$, are needed, where $d\gamma^{(k)}$ is the shear strain on the kth slip system of a crystal. The ratio of the sum of shear strains on active slip systems to the plastic strain increment is often called the Taylor factor (*Tf*) [71], which is expressed as

$$Tf = \sum d\gamma^{(k)} / d\varepsilon_{ij} \quad (33)$$

The *Tf* predicts that the strain energy of grains increases with increasing their *Tf*s. We suppose that the present Cu deposit and its grains undergo the same equibiaxial plastic strain. This strain state is equivalent to an axisymmetric strain state. Chin and Mammel [65] calculated orientation dependence of *Tf* for axisymmetric deformation for {110}<111> or {111}<110> slip systems. *Tfs* of the <100>//ND, <112>//ND, <110>//ND, and <111>//ND fcc crystals are 2.449, 3.076, 3.674, and 3.674, respectively. *Tf* of the <100>//ND crystal is the minimum and the <110>//ND and <111>//ND crystals have the maximum *Tf*. This means that the <100>//ND grains have the minimum strain energy, and the <110> and <111>//ND grains have the maximum strain energy. Therefore, the <100>//ND grains seems likely to grow at the expense of neighboring grains with higher *Tfs*.

However, all grains in the Cu deposit cannot undergo the same strain as the grain aggregate (deposit) does. The thermal strain increases with increasing annealing temperature. During that course, some grains will undergo plastic deformation before others, and the plastically deformed grains are likely to have higher strain energy due to dislocations generated by plastic deformation than surrounding grains with different orientations, which still remain in the elastic regime. In fact, an increase in *Tf* also means an increase in resistance to deformation and thus *Tf* can be expressed as [72,73]

$$Tf = \sum d\gamma^{(k)} / d\varepsilon_{ij} = \sigma_{ij} / \tau_c \tag{34}$$

where σ_{ij} is the strength and τ_c is the critical shear stress acting on all the active slip system. For instance, grains which are oriented to have infinite *Tf* cannot be plastically deformed. When a polycrystalline aggregate is subjected to a plastic deformation that is not so high enough to make the strengths of grains become similar, low *Tf* grains are likely to deform more than neighboring high *Tf* grains. In this case, low *Tf* grains have higher strain energy than high *Tf* grains, which in turn makes high-*Tf* grains grow at the expense of neighboring low *Tf* grains. These phenomena have been found in annealing textures of Ag [66,67], Cu-bearing bake hardening steel [68], and Al [69].

For convenience's sake, we consider <100>//ND, <112>//ND, <110>//ND, and <111>//ND grains. The <100>//ND grains are likely to begin to be deformed before grains with other orientations because the <100>//ND grain has the lowest *Tf*. For the increased strain (1.39×10^{-3}) at 125°C, we expect that grains with orientations of lower *Tfs* begin to undergo plastic deformation. The evolution of the strong <112>//ND texture after annealing at 125, 300, and 300°C (Figures 30 to 32, Table 4) can be explained by assuming that <100>//ND grains with the lowest *Tf* have been deformed at these temperatures, while grains with the other three orientations remain in the elastic regime. In this case, the <112> grains would grow at the expense of the <110>//ND and <111>//ND grains in the elastic regime. Simultaneously, they are likely to grow by consuming the <100> grains in the plastic regime. As a result, the growth of <112>//ND grains would be favored by elastic strain energy minimization, replacing the textures found in the as-deposited film.

After annealing at 125°C, the area fractions of the <112>//ND component did not vary between the matrix and AGG textures (Figure 30 b), with those of the <100>//ND and <111>//ND grains even increased, whereas the area fraction of the <112>//ND component in the AGG texture was observed to be much higher than that in the matrix texture at 300°C (Figure 31 b-c). It has been suggested that at 125°C, parts of the <100>//ND grains are still in the elastic regime, contributing to AGG. And at the temperature, strain energy minimization seemed not so completely dominant over surface energy minimization. As mentioned above, strain energy anisotropy is likely to be stronger with increasing temperature further. Thus at 300°C, strain energy minimization would be controlling over surface energy minimization and due to a steady buildup of thermal strain, more <100>//ND grains are expected to belong to the plastic regime. Under such circumstance, when AGG occurred, the <112> grains would make up a dominant area fraction, as shown in Figure 31 b and c. After annealing at 500°C, the <111> component was observed to make up the largest area fraction with the area fraction of the <112>//ND component still comparable (Figure 32 c). The evolution of the <111>//ND component is naturally attributed to surface energy minimization. At such a high temperature, defects inside grains gradually diminish and the surface energy increasingly dominates because the atomic mobility increases with increasing temperature, enabling <111>//ND grains to grow abnormally.

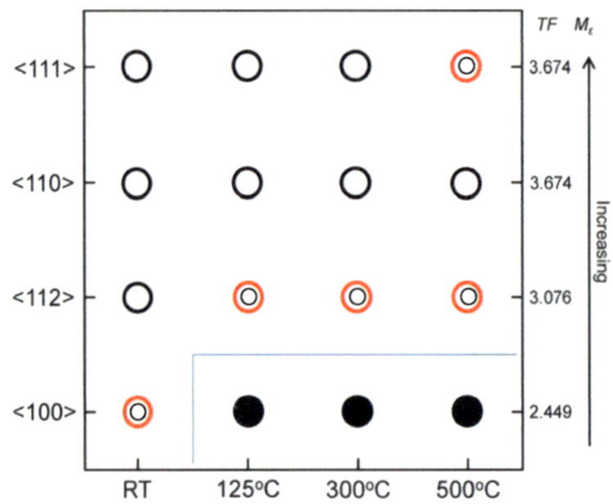

Figure 33. Schematic diagram showing that nanocrystalline grains in film system under plane stress, equibiaxial strain state undergo elastic–plastic transformation with increasing temperature, transition which is closely associated with *Tf*. Open circles indicate grain orientations in elastic regime, while closed ones are in plastic regime. AGG texture is suggested to be dominated by grains with lowest biaxial elastic modulus (M_ε) in elastic regime at each temperature, as indicated by the double circles. An exception seems to occur at 500°C, where surface energy minimization becomes dominant, leading to major <111> texture component [59].

Taken together, the evolution of the <100>//ND component during AGG for the as-deposited film is purely explained by elastic strain energy minimization. The <111> component in the AGG textures for the as-deposited film and the film annealed at 500°C is explained by surface energy minimization. The development of the <112> texture for the deposits annealed at 125, 300, and 500°C is made clear by assuming that <100>//ND grains were plastically deformed at these temperatures. Out of grains which have higher *Tf*s and thus are still in the elastic regime, those with the lowest biaxial elastic modulus (i.e., <112>//ND) would grow at the expense of the surrounding grains. This model can apply to the development of <112>//ND during AGG in the <111>//ND matrix Al alloy thin films reported by Longworth and Thompson [70]. These results are summarized in Figure 33.

http://dx.doi.org/10.5772/59724

This study [59] opens a way to the development of a comprehensive mechanism for the AGG texture evolution in thin films taking two factors into consideration: strain energy minimization and *Tfs*. If all grains belong to the elastic regime, the AGG texture is simply determined by elastic strain energy minimization, which explains the development of the AGG texture of <100>//ND found in the as-deposit deposited film. As a thermal strain builds up with increasing annealing temperature above room temperature, grains with lower *Tfs* in thin films will undergo plastic deformation before other grains with higher *Tfs*. In this case, grains undergoing plastic deformation would be consumed by grains still remaining in the elastic regime. Among grains in the elastic regime, grains with the minimum biaxial elastic modulus will grow at the expense of others. As a result, grains with the minimum elastic biaxial modulus out of grains in the elastic regime grow abnormally. The hypothesis well clarifies the evolution of <112> as abnormal grains during annealing at 125 and 300°C.

6. Conclusion

Silver and copper wires subjected to heavy cold-drawing evolve the <111>//AD texture, where AD stands for the axial direction. When recrystallized, the <111>//AD drawing texture turns to the recrystallization texture composed of major <100>//AD and minor <111>//AD. As annealing time prolonged, the <111>//AD component increased, while the <100> component decreases, accompanied by abnormal grain growth. The texture change during abnormal grain growth must be related to grain boundary mobility because the dislocation density drastically decreases during the primary recrystallization and the internal energy of the system is dominated by the grain boundary energy.

The textures (~ major <100>//ND + minor <111>//ND) of free standing nanocrystalline Ni and Fe-Ni alloy electrodeposits of 20 to 30 μm in thickness and in the order of 10 nm in grain size change to the major <111> + minor <100>//ND texture after annealing, where ND indicates the deposit-surface normal direction. The most important factor dominating the texture change is the grain boundary mobility. The texture transition mechanism is similar to that in abnormal grain growth textures in drawn silver and copper wires. Another factor is the <111>//ND grains being purer than the <100>//ND grains.

The structure of electroless Ni-Co-P alloy deposits of 30 to 40 μm in thickness on a 5086 aluminum alloy sheet is characterized by 6- to 7-nm-sized crystallites imbedded in amorphous matrix. The (111) peak intensities of the deposits are much higher than the (200) intensities. When annealed for 2 h, the (200) peak intensities of the deposits increase more rapidly than the (111) intensity with increasing annealing temperature and cobalt content in the deposits [24]. At an annealing temperature of 325°C, Ni_5P_2 forms especially in the higher phosphorous deposits, but it disappears and stable Ni_3P appeared at higher temperatures [24]. The evolution of the <100>//ND texture in fcc metal films with the <111>//ND texture on a substrate during annealing is due to the fact that the <100>//ND grains have lower thermal strain energy than the <111>//ND grains.

When Al interconnects made by patterning of 500-nm-thick Al-1%Cu films of 500 nm deposited over Ti and TiN sublayers onto a Si (001) are annealed at 460°C, a near-bamboo character is obtained for those lines having widths less than 2.0 μm and a polycrystalline structure was obtained for the wider lines. The resultant grain size is of the order of 1 μm and varies in proportion to the line width. For narrow, bamboo-structured lines align with the <110> direction. The <111>//ND texture strength increased with decreasing line width during annealing. The result is explained by the surface energy minimization of the system.

The initial size of Cu grains in Cu interconnects of 700 nm in thickness made by electroplating Cu into damascene trenches of various width/space combinations is not larger than 15 nm. The growth rate of Cu grains is faster in the pattern area than in the blanket area. The initiation of the grain growth occurs at the upper corners of the trench plugs. The number density of corners is higher in the pattern area than in the blanket area. The textures of all the specimens except the blanket specimen can be approximated by the {111}<110> texture as a major component and its twin components {115}<110> and {115}<141> as minor ones. The texture of the deposit on blanket wafer was approximated by a <111>//ND texture, which inherited from Cu seed layer. The self-annealing textures of the 200 nm/200 nm and 240 nm/200 nm width/space patterned trenches were approximated by a major {111}<110> component and a minor {115}<110> and {115}<141> components, whereas the texture of the 1000 nm/200 nm width/space specimen was approximated by a major {111}<112> component and a minor {115}<552> component, which is a twin component of {111}<112>. The textures of the 500 nm/200 nm, 2000 nm/200 nm, 500 nm/500 nm, and 500 nm/1000 nm specimens are seen to be mixtures of the {111}<110> and {111}<112> orientations, contributions of which vary with specimen. The 4- and 6-μm-wide specimens showed textures which could be approximated by mixtures of diffused {111}<110> and {111}<112> components. The textures measured over the over-plated layer were similar to those of the CMP-specimens. The annealing texture is determined by the strain energy minimization.

The overall and matrix textures of nanocrystalline Cu films deposited by DC sputtering at room temperature onto amorphous SiN_x/Si TEM grids with a viewable area of 100x100 μm^2 are dominated by <110>//ND, but <100> //ND grains and a smaller fraction of <111>//ND grains grow abnormally by elastic strain energy minimization and by surface energy minimization. After annealing at 125 and 300°C, AGG develops strong <112>//ND component by the <100>// ND grains undergoing plastic deformation. Out of grains which have higher *Tf*s and are still in the elastic regime, those with the lowest biaxial elastic modulus (i.e., <112>//ND) would grow at the expense of the surrounding grains. At 500°C, the major AGG texture component becomes <111>//ND by surface energy minimization.

Acknowledgements

This study was supported by the Basic Science Research Program through the National Research Foundation of Korea (NRF) funded by the Ministry of Education (NRF-2013R1A1A2005181) and RIAM.

Author details

Dong Nyung Lee* and Sung Bo Lee*

*Address all correspondence to: dnlee@snu.ac.kr; bolee@snu.ac.kr

Department of Materials Science and Engineering, and RIAM, Seoul National University, Seoul, Republic of Korea

References

[1] Lee DN, Han HN. Recrystallization textures of metals and alloys. In: Wilson P. (ed). Materials Science. InTech; 2013. http://www.intechopen.com/articles/show/title/recrystallization-textures-of-metals-and-alloys

[2] Lee DN, Lee H-J. Effect of stresses on the evolution of annealing textures in Cu and Al interconnects. Journal of Electronic Materials 2003;32 1012-1022.

[3] Shin H-J, Jeong H-T, Lee DN. Deformation and annealing textures of silver wire. Materials Science and Engineering 2000;A279 244-253.

[4] Abbruzzese G, Lücke K. A theory of texture controlled grain growth-I. Derivation and general discussion of the model. Acta Metallurgica 1986;34 905-914.

[5] Hillert M. On the theory of normal and abnormal grain growth. Acta Metallurgica 1965;13 227-238.

[6] Abbruzzese G. Computer simulated grain growth stagnation. Acta Metallurgica 1985;33 1329-1337.

[7] Eichelkraut H, Abbruzzese G, Lücke K. A theory of texture controlled grain growth-II. Numerical and analytical treatment of grain growth in the presence of two texture components. Acta Metallurgica 1988;36 55-68.

[8] Aleshin AN, Aristov VY, Bokstein BS, Shvindlerman LS. Kinetic properties of <111> tilt boundaries in aluminium. Physica Status Solidi 1978;A45 359-366.

[9] Wolf D. Structure-energy correlation for grain boundaries in F.C.C. metals-III. Symmetrical tilt boundaries. Acta Metallurgica et Materialia 1990;38 781-790.

[10] Grant E, Juul Jensen D, Ralph B, Hansen N. Texture development in pure copper. In: Brakman CM, Jongenburger P, Mittemeijer EJ (eds.). Seventh International Conference on Textures of Materials: proceedings of ICOTOM 7. Noordwijkerhout, The Netherlands: Netherlands Society for Materials Science; 1984.

[11] Inakazu N, Kaneno Y, Inoue H. Fiber texture formation and mechanical properties in drawn fine copper wire. Materials Science Forum 1994;157-162 715-720.

[12] Humphreys FJ, Hatherly M. Recrystallization and Related Annealing Phenomena. Pergamon Press; 1995.

[13] Park H, Lee DN. The evolution of annealing textures in 90 pct drawn copper wire. Metallurgical and Materials Transactions 2003;34A 531-541.

[14] Shvindlerman LS, Gottstein G, Molodov DA, Sursaeva VG. Triple junction motion in metals. In: Gottstein G, Molodov DA (eds.), Recrystallization and Grain Growth: proceedings of The First Joint International Conference on Recrystallization and Grain Growth. RWTH Aachen, Germany: Springer-Verlag; 2001.

[15] Porter DA, Eastering KE. Phase Transformations in Metals and Alloys. London: Champman & Hall; 1992.

[16] Harase J. Shimizu R. Dingley DJ. Texture evolution in the presence of precipitates in Fe–3% Si alloy. Acta Metallurgica et Materialia 1991;39 763-770.

[17] Doherty RD, Hughes DA, Humphreys FJ, Jonas JJ, Juul Jensen D, Kassner ME, King WE, McNelley TR, McQeen HJ, Rollett AD. Current issues in recrystallization: a review. Materials Science Engineering 1997;A238 219-274.

[18] Hayakawa Y, Muraki M, Szpunar JA. The changes of grain boundary character distribution during the secondary recrystallization of electrical steel. Acta Metallurgica 1998;46 1063-1073.

[19] Gottstein G, Shvindlerman LS. Grain Boundary Migration in Metals, Boca Raton: CRC Press; 1999. p. 203.

[20] Molodov DA. Grain boundary character – a key factor for grain boundary control. In: Gottstein G, Molodov DA (eds.), Recrystallization and Grain Growth: proceedings of The First Joint International Conference on Recrystallization and Grain Growth. RWTH Aachen, Germany: Springer-Verlag; 2001.

[21] Czerwinski F, Li H, Megret M, Szpunar JA, Clark DG, Erb U. The evolution of texture and grain size during annealing of nanocrystalline Ni-45%Fe electrodeposits. Scripta Materialia 1997; 37 1967-1972.

[22] Park YB, Park J, Ha CS, Yim TH. Texture evolution during the annealing of nanocrystalline permalloy. Materials Science Forum 2002;408-412 919-924.

[23] Park YB, Hong S-H, Ha CS, Lee HY, Yim TH. Orientation dependence of grain growth in a nanocrystalline Ni foil. Materials Science Forum 2002;408-412 931-936.

[24] Lee DN, Hur KH. The evolution of texture during annealing of electroless Ni-Co-P deposits. Scripta Materialia 1999;40 1333-1339.

[25] Lee DN, Hur KH. The evolution of texture during annealing of nanocrystalline electroless Ni alloy deposits. Textures and Microstructures 2000;34 181-195.

[26] Lee DN. Changes in lattice constants and orientation of <100> and <111> oriented grains in nanocrystalline Ni and Ni-Fe electrodeposits after annealing. Materials Science Forum 2007;539-543 149-154.

[27] Metals Handbook. American Society for Metals. 9th edition, Vol. 1, 1979. p.147.

[28] Metals Handbook. American Society for Metals. 9th edition, Vol. 2, 1979. p. 105 for 5086 Al alloy, p. 777 for Ni, p.727 For Cu.

[29] Barrett CS, Massalski TB. Structure of Metals, 3rd edition, McGraw-Hill Book Company; 1966. p. 629.

[30] Metals Handbook. American Society for Metals. 8th edition, Vol. 1, p. 1207.

[31] Hur K-H, Jeong J-H, Lee DN. Effect of annealing on magnetic properties and microstructure of electroless nickel-copper-phosphorous alloy deposits. Journal of Materials Science 1991;26 2037-2044.

[32] Lee DN. Annealing textures of thin films and copper interconnects. Materials Science Forum 2005;475-479 1-8.

[33] Lee DN, Kang S, Yang J. Relation between texture and surface morphology of copper electrodeposits. Plating and Surface Finishing March 1995; 82 76-79.

[34] Carel R, Thompson CV, Frost HJ. Computer simulation of strain energy effects vs surface and interface energy effects on grain growth in thin films. Acta Materialia 1996;44 2479-2494.

[35] Lee DN. Elastic properties of thin films of cubic system. Thin Solid Films 2003;434 183-189.

[36] Lee DN. Corrigendum to [Thin Solid Films 434, 183-189 (2003)], Thin Solid Films 2012;520 3708.

[37] Hearmon RFS. The elastic constants of anisotropic materials—II. Advances in Physics 1956;5 323-382.

[38] Field DP, Sanchez, Jr. JE, Besser PR, Dingley DJ. Analysis of grain-boundary structure in Al–Cu interconnects. Journal of Applied Physics 1997;82 2383-2392.

[39] Field DP, Wang P-H. Development of a preferred grain boundary structure in Al–Cu Interconnects. In: Pond RC, Clark WAT, King AH, Williams DB (eds). Boundaries and Interfaces in Materials: The David A. Smith Symposium, The Minerals, Metals & Materials Society. 1998. pp. 323-328.

[40] Sundquist BE. A direct determination of the anisotropy of the surface free energy of solid gold, silver, copper, nickel, and alpha and gamma iron. Acta Metallurgica et Materialia 1964;12 67-86.

[41] Lee H-J, Kim DI, Ahn JH, Lee DN. Electron backscattered diffraction analysis of copper damascene interconnect for ultralarge-scale integration. Thin Solid Films 2005;474 250-254.

[42] Lingk C, Gross ME. Recrystallization kinetics of electroplated Cu in damascene trenches at room temperature. Journal of Applied Physics 1998;84 5547-5553.

[43] Lee DN, Lee H-J. Self-annealing textures of copper damascene interconnects, Materials Science Forum 2004;467-460 1333-1338.

[44] Lee H-J, Han HN, Lee DN. Annealing texture of copper interconnects for ultralarge scale integration, Journal of Electronic Materials 2005;34 1493-1499.

[45] Kwon K-W, Ryu C, Sinclair R, Wong SS. Evidence of heteroepitaxial growth of copper on beta-tantalum. Applied Physics Letters 1997;71 3069-3071.

[46] Lee DN. Texture development in thin films, Materials Science Forum 2002;408-412 75-94.

[47] Lee DN. Current understanding of annealing texture evolution in thin films and interconnects. Zeitschrift für Metalkunde 2005;96 259-268.

[48] Brongersma SH, Kerr E, Vervoort I, Maex K. Limitations to copper grain growth in narrow trenches. Proc. IEEE Int. Interconnect Technology Conference; 2001. p. 230.

[49] Ueno K, Ritzdorf T, Grace S. Seed layer dependence of room-temperature recrystallization in electroplated copper films. Journal of Applied Physics 1999;86 4930-4935.

[50] Lagrange S, Brongersma SH, Judelewicz M, Saerens A, Vervoort L, Richard E, Palmans R, Maex K. Self-annealing characterization of electroplated copper films. Microelectronic Engineering 2000;50 449-457.

[51] Teh WH, Koh LT, Chen SM, Xie J, Li CY, Foo PD. Study of microstructure and resistivity evolution for electroplated copper films at near-room temperature. Microelectronics Journal 2001;32 579-585.

[52] Reeber RR, Wang K. Thermal expansion and lattice parameters of group IV semiconductors. Materials Chemistry and Physics 1996;46, 259-264.

[53] Hellwege KH, Hellwege AM (eds.). Landolt-Börnstein: Numerical Data and Functional Relationships in Science and Technology. Group-III, vol.1. New York: Springer; 1966.

[54] Zhao J-H, Ryan T, Ho PS, Mckerrow AJ, Shih W-Y. Measurement of elastic modulus, Poisson ratio, and coefficient of thermal expansion of on-wafer submicron films. Journal of Applied Physics 1999;85 6421-6424.

[55] Zhao J-H, Du Y, Morgen M, Ho PS. Simultaneous measurement of Young's modulus, Poisson ratio, and coefficient of thermal expansion of thin films on substrates. Journal of Applied Physics 2000;87 1575-1577.

[56] Brandes EA, Brook GB (eds.). Smithells Metals Reference Book. 7th edition, Butterworth-Heinemann; 1999.

[57] Lynch CT (ed.). CRC Handbook of Materials Science. vol. II, Boca Raton, FL: CRC Press; 1975.

[58] Shen Y-L, Suresh S, Blech IA. Stresses, curvatures, and shape changes arising from patterned lines on silicon wafers. Journal of Applied Physics 1996;80 1388-1398.

[59] Lee SB, Kim D-I, Hong S-H, Lee DN. Texture evolution of abnormal grains with post-deposition annealing temperature in nanocrystalline Cu thin films. Metallurgical and Materials Transactions A 2013;44 152-162.

[60] Lee DN. A model for development of orientation of vapor deposits. Journal of Materials Science 1989;24 4375-4378.

[61] Lee DN. Textures and structures of vapor deposits. Journal of Materials Science 1999;34 2575-2582.

[62] Kwatera A. Thin CVD layers of carbon-doped silicon nitride on quartz. Ceramics International 1989;15 65-129.

[63] Chang YA, Himmel L. Temperature dependence of the elastic constants of Cu, Ag, and Au. Journal of Applied Physics 1966;37 3567-3572.

[64] Detavernier C, Deduytsch D, Van Meirhaeghe RL, DeBaerdemaeker J, Dauwe C. Room-temperature grain growth in sputter-deposited Cu films. Applied Physics Letters 2003;82 1863-1865.

[65] Chin GY, Mammel WL. Computer solutions of the Taylor analysis for axisymmetric flow. Transactions of the Metallurgical Society of AIME 1967;239 1400-1405.

[66] Lee DN, Han HN, Kim S-J. Rolling and annealing textures of silver sheets. Proceedings of ICOTOM 15, Symposium 10, TMS, Pittsburgh; 2008.

[67] Kim S-J, Han HN, Jeong H-T, Lee DN. Evolution of the {110}<110> texture in silver sheets. Materials Research. Innovations 2011;15 (Suppl.1) s390-s394.

[68] Oh KH, Park SM, Koo YM, Lee DN. Thermomechanical treatment for enhancing gamma fiber component in recrystallization texture of copper-bearing bake hardening steel. Materials Science and Engineering 2011;A528 6455-6462.

[69] Merriman CC, Field DP, Trivedi P. Orientation dependence of dislocation structure evolution during cold rolling of aluminum. Materials Science and Engineering 2008;A494 28-35.

[70] Longworth HP, Thompson CV. Abnormal grain growth in aluminum alloy thin films. Journal of Applied Physics 1991;69 3929-3940.

[71] Taylor GI. Plastic strain in metals. Journal of the Institute of Metals 1938;62 307-324.

[72] Bishop JFW, Hill R. A theory of the plastic distortion of a polycrystalline aggregate under combined stresses. Philosophical Magazine 1951;42 414-427.

[73] Bishop JFW, Hill R. Atheoretical derivation of the plastic properties of polycrystalline face-centered metal. Philosophical Magazine 1951;42 1298-1307.

Chapter 2

Physical and Technological Aspects of Processing High-Purity Refractory Metals

Vadim Glebovsky

Additional information is available at the end of the chapter

http://dx.doi.org/10.5772/59397

1. Introduction

The chapter presented is in some sense autobiographical. At one time during my experiments, I was faced with a huge obstacle: to get high-melting metals, namely, molybdenum, tungsten, and niobium, of very high purity. The high purity of refractory metals primarily means the low content of the two stubborn interstitials – carbon and oxygen. The presence of carbon and oxygen in polycrystalline metals melted by an electron beam in a vacuum often makes further plastic deformation of the metal difficult and unpredictable due to increased brittleness. In part, –what is clear is that there is a segregation of interstitial impurities at grain boundaries, which complicates plastic deformation. The literature on this subject points to a similar conclusion, with a universal solution: to eliminate effects of segregation. This path is well tested during the steelmaking through the introduction of deoxidants and modifiers, which include a variety of chemical elements, including carbon. Our objective is to obtain the "elemental" metal, in which impurities must not exceed a few tens of ppb. The question then arises: To what extent can these impurities be removed? Is it even possible to obtain the refractory metal with an extremely low content of carbon and oxygen? What is the mechanism of the removal of interstitial impurities? What analytical methods can control the removal processes with an adequate accuracy? It turned out that these questions, to a large extent, do not have a reliable answer. Moreover, experiments aimed at the solution of similar problems have been very inconsistent, so we have to carry out a detailed analysis of the main theoretical and experimental works on this topic. The analysis has been done, and the theoretical model of the behavior of carbon and oxygen in the vacuum annealing of solid refractory metals, as well as of the special case – of the interaction of these interstitials in the liquid refractory metals in vacuum – has been developed. Knowing the complicated behavior of carbon and oxygen in refractory metals, experiments are conducted using crucible-free melting methods – vacuum levitation of small droplets (15–20 g) in a high-frequency electromagnetic field and electron beam zone melting with a floating zone of the molten metal. Both methods are non-contact in nature and have a high rate of convection of the melt, that is, the liquid metal in the both methods has no contact with any of the refractory material, and the volume of the liquid metal contains convective flows acting at a high rate. This makes it possible to perform experiments actually in the diffusion-less area when the equilibrium of the carbon–oxygen reaction in a bulk of the liquid droplet can be achieved in seconds. In addition, for controlling the oxygen and carbon, the highly sensitive methods of analysis (e.g., the fast neutron activation analysis of oxygen and deuterons for the analysis of carbon) have been used. All this make it possible to establish and prove that under certain conditions (mainly the adequate initial content of carbon and oxygen), as a result of the interaction, both impurities can be arbitrarily very low. This knowledge is used when performing the second task: obtaining ingots of high-purity molybdenum and tungsten. Technological options for obtaining massive ingots of high-purity unalloyed molybdenum and tungsten of the high technological plasticity during plastic deformation have been developed. The rolled products are obtained in the form of the thin sheets and foils of high quality. This chapter presents consistent data to give a reader an idea of the present study of the interaction of carbon and oxygen in refractory metals and the essence of the developed theoretical model. For solving the puzzle of low carbon and oxygen, the first half of this chapter can be utilized. At least, if a reader would like to solve puzzles, he can turn the reading of this Chapter, especially the first half of it, in a very entertaining lesson. A separate section is devoted to the experimental confirmation of the validity of the developed model using crucible-free methods of melting and highly sensitive methods for analyzing the content of impurities in the samples of both molybdenum and niobium. In the last section of the chapter, the experimental data are presented on the technological plasticity of large ingots of molybdenum and tungsten (up to hundreds of kilograms).

2. Theoretical and experimental aspects of obtaining pure refractory metals

2.1. Characteristics of gas-forming impurities removal

The removal of interstitial impurities dissolved in the metal, in the gas phase or vacuum, is a complex process consisting of the following elementary steps [1]:

1. Diffusion transition of impurities from the volume of metal to the surface layer: $A_{dis}^{v} \rightarrow A_{dis}^{s}$.

2. Transition of impurity atoms across the interface on the metal surface in the chemisorbed state: $A_{dis}^{s} \rightarrow A_{ads}$.

3. Recombination of adsorbed atoms (adatoms) with adsorbed particles (adparticles) coming to the surface of the metal or to the gas phase.

4. Desorption of adsorbed molecules or direct desorption of adatoms impurities in the gas phase or vacuum: $A_{ads} \rightarrow A_{gas}$ or $A_{ads} + B_{ads} (B_{gas}) \rightarrow AB_{gas}$. A kinetic analysis of the complex

heterogeneous process has been conducted on the basis of the system of equations obtained from the conservation of the particle number. Assuming that the exchange of the impurity atoms between the surface and the gas phase (vacuum) is carried out by the desorption and adsorption flows with densities J^a and J^d, and between the surface and the subsurface layer, by the flows with densities J^θ and J^s, we obtain:

$$\partial N / \partial t = D\nabla^2 N \tag{1}$$

$$dN^n\big|_s /dt = J^D\big|_s - J^{s\theta} \tag{2}$$

$$n / dt = -J^d + J^{s\theta} \tag{3}$$

where N is the bulk impurity content, $dN^n|_s$ the planar impurity content in the surface layer, n the impurity content of the adatoms on the surface, $J^{s\theta}$ the resultant flux density from the subsurface layer to the surface, $J^D|_s = -D/[v_a(\nabla N)_s]$ the diffusion flux density in the interface, D the diffusion coefficient of impurities, v_a the atomic volume of metal–matrix (the content of impurity atoms is expressed in atomic fractions).

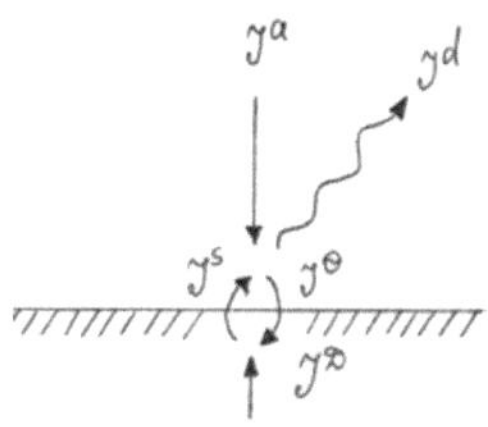

The system of equations can be solved only by numerical methods because of its complexity. The substantial simplification can be achieved on the basis of a quasi-stationary approximation, according to which the impurity flows quickly adapt to each other so that the change in the content of impurities on the surface and in the subsurface layer is insignificant and can be neglected (a quasi-steady state). Assuming

$$J^D\big|_s - J^{s\theta} = dN^{\ddot{\imath}}\big|_s / dt \simeq 0; J^a - J^d + J^{s\theta} = dn / dt \simeq 0.$$

For two flux equations we write the matching condition of fluxes:

$$J^D\big|_s = \frac{-D}{v_a(\nabla N)_s} = J^{s\theta} = J^d - J^a. \tag{4}$$

Equalities (4) make it possible to formulate the boundary conditions within the framework of the adopted model to the diffusion equation (1) and determine them together with the boundary value problem. A solution to this problem allows the determination of the nature of changes in a local content of impurity atoms $N(\vec{r}, t)$, an average content of impurity atoms for a sample $N(t)>\int N(\vec{r}, t)dv$ and surface content of impurity atoms $n(t)$ in the course of their redistribution between the volume of metal, the surface, and the gas phase (vacuum).

2.2. Differences in energy state

Differences in the energy states of the impurity atom on the surface and in the metal lattice lead to an asymmetry of the potential relief in the vicinity of the interface (Figure 1). They lead to the well-known phenomenon of surface segregation of impurities [2-4]. Here, for a description of this phenomenon, reacting adatoms of impurities on the surface is usually neglected, and the surface energy is considered homogeneous. Within Langmuir approach for flux densities J^s and J^θ [5–8]:

$$J^s = v^s N|_s (1-\theta) \tag{5}$$

$$J^\theta = \frac{1-C_s}{v^\theta \theta} \tag{6}$$

where, $\theta = n / n_m$ is the degree of filling of the adsorption sites on the surface, $C = N/N_m$ the degree of filling positions in the solution implementation, v^s and v^θ the constants of the rate, which are proportional, respectively, to transition probabilities of impurity atoms from the subsurface layer to the surface and back. Thus, the flux density from the subsurface layer to the surface can be written as

$$J^{s\theta} = v^s N_s (1-\theta) - v^\theta \theta (1 - C|_s), \tag{7}$$

or for the area of a dilute solution when $C|_s \ll 1, \theta \ll 1$

$$J^{s\theta} = v^s N|_s - v^\theta \theta. \tag{8}$$

An exchange of impurity atoms between the surface and the subsurface layer ($J^s \rightleftharpoons J^\theta$) occurs very rapidly, and at this stage of the quasi-equilibrium is established: $J^s \cong J^\theta \gg J^{s\theta}$.

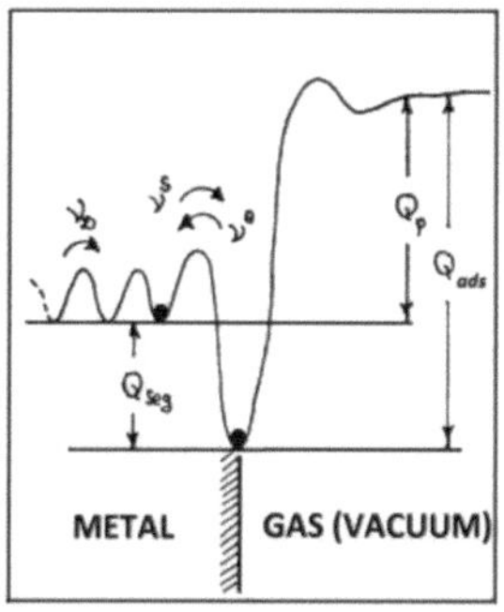

Note: Q_{seg} is the heat of segregation, Q_s the heat of dissolution, and Q_{ads} the heat of adsorption.

Figure 1. The potential relief for impurity atom in the vicinity of the metal–gas interface.

2.3. Adsorption and desorption phenomena on the surface of metals

Depending on the nature and strength of the bond of the adatoms and the admolecules the metal adsorption can be of two types: physical and chemical (chemisorption). In a state of physical adsorption, the adparticles hold on the metal surface by weak van der Waals forces and retain their individuality by having only a small perturbation of the metal. In this case, the physical adsorption heat is typically closer to the heat of molecular condensation and usually does not exceed 10 kcal mol^{-1}. In chemisorption the adparticles form strong chemical bonds (via an electronic exchange) with the metal surface, which in one way or another include covalent, ionic, or "metal" components and have been localized near certain active centers, distributed in accordance with the geometry of the surface structure of a solid metal. The heat of chemisorption is close to the heat of chemical reactions and often reaches several hundred kilocalories [9-12]. Typical lifetimes of adsorbed particles on the surface of a metal depending on their energy due to the metal can be estimated using the Frenkel equation [11]:

$$\tau = \tau_0 exp\left(Q / RT\right), \tag{9}$$

where τ_0 is the molecular vibration period equal to ~10^{-13} s; Q the heat of adsorption. Average lifetimes of adsorption of particles, calculated on the equation for certain values of Q and T, are shown in Table 1. It can be seen that a physically adsorbed film even at room temperature flies from the surface of the metal almost instantly, while the particles that are not strongly chemisorbed desorb even at temperatures above 1000 K.

A strong chemisorption is characteristic of interstitial impurities (hydrogen, oxygen, nitrogen, carbon), with a high "valence activity," which are adsorbed on the surface of transition refractory metals with relatively directed α-orbitals. By virtue of this, a direct desorption of the impurities is only noticeable at very high temperatures of the solid metal. In this case, in accordance with the first order reaction $A_{ads} \rightarrow A_{gas}$

Binding energy Q, kcal.mole^{-1}	Average lifetimes of adparticles τ, s		Bond
300K	1000K		
<10	<10^{-6}	<10^{-11}	Physical adsorption
20	~30	~10^{-9}	Weak chemisorption
50	~10^{23}	~10^{-2}	Middle chemisorption
>100	>10^{60}	>10^{8}	Strong chemisorption

Table 1. Average lifetimes of adparticles calculated for certain values of the adsorption heat and temperature

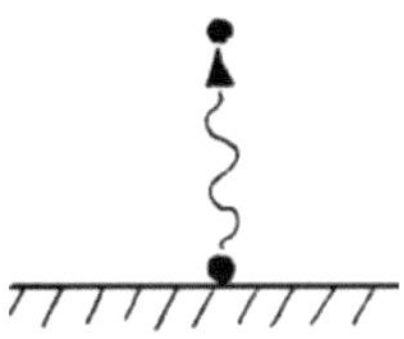

the desorption flux density is proportional to the content of adatoms [9,13]:

$$J_d = k_\alpha \theta \simeq n_m \theta / \tau, \tag{10}$$

where n_m is the density of available adsorption sites, which is ~10^{15} at cm^{-2}. Despite the fact that the mechanism of the adatoms transition in the gas phase is the most simplest, it dominates by a relatively rare, as most energetically preferred, desorption processes of adatoms as a result of their recombination with other adsorbed particles presenting on the surface according with Langmuir-Hinshelwood mechanism, $A_{ads} + B_{ads} \rightarrow AB_{gas}$ (I) or as a result of their recombination with molecules impinging from the gas phase according with Ili-Ridil mechanism, $A_{ads} + B_{gas} \rightarrow AB_{gas}$ (II):

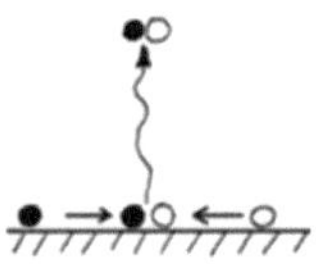
(I)

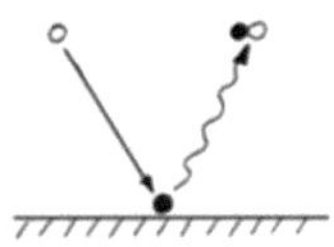
(II)

At the high vacuum annealing, the probability of an interaction with the molecules of the residual gas phase is relatively small, so the determining role belongs to the recombination processes by Langmuir mechanism (I). The energy required for breaking the bonds M–A_{ads} and M–B_{ads} (respectively Q_A and Q_B) is partly compensated by a heat release of the dissociation (D_{AB}) of AB_{gas} molecules ($A_{gas} + B_{gas} \rightarrow AB_{gas} + D_{AB}$). An effective activation energy, which at a weakly activated nature of the adsorption of molecules AB, is close to the heat of adsorption is: $E'_{d,AB} \simeq Q_{AB} = Q_A + Q_B - D_{AB}$. By comparing magnitudes of $E'_{d,AB}$ and Q_A the relatively simple

http://dx.doi.org/10.5772/59397

energy criterion can be achieved, according to which the desorption of adatoms A_{ads} by the reaction $A_{ads} + B_{ads} \rightarrow AB_{gas}$ would dominate their desorption in the atomic state in the case when the binding energy of the particles B with the metal is less than their binding energy in the molecule AB_{gas}:

$$Q_B < D_{AB} \tag{11}$$

In other words, the active partners of the recombination are only those particles that have a large enough "chemical affinity" with respect to the atoms of A and comparatively weak hold on the metal. In cases where the surface of the metal adatoms are only of one type A or there are relatively few active adsorption particles of any other type B, the direct desorption of adatoms can compete with two mechanisms, which are special cases of the process (I):

1. The pair recombination of adatoms of one type followed by the molecules desorption of simple diatomic gases $A_{ads}+A_{ads} \rightarrow A_{2gas}$ (III):

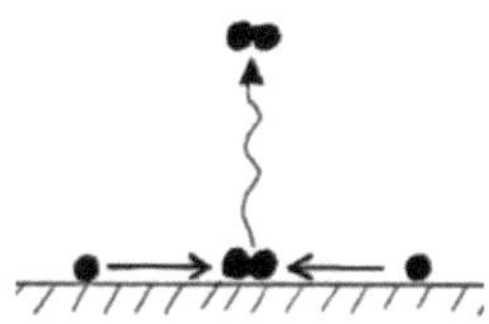

2. The formation and desorption of the compounds with the metal atoms of the substrate have the form M_nA_m, and in the simplest cases, MA: $A_{ads}+ M \rightarrow MA_v$ (IV).

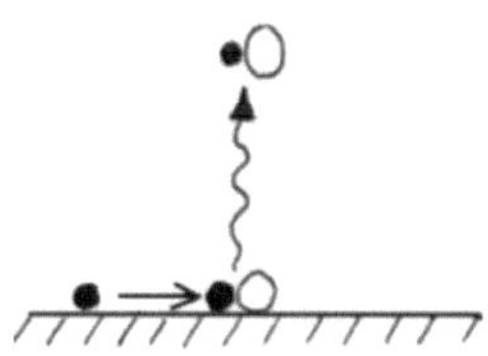

The condition (11) in the first case restricts the binding energy of the adatom–metal substrate [9]:

$$Q_A < D_{A2} \tag{12}$$

In a second case, the condition (11) limits an amount of the sublimation heat of the metal substrate (L_M) [14,15]:

$$L_M < D_{MA} \tag{13}$$

Thus, at a relatively weak bond of adatoms with the metal substrate ($Q_A < \min\{D_A, L_M - D_{MA} + D_{A2}\}$) the reversible selection of A_{ads} in the molecular state ($2A_{ads} \rightleftharpoons A2_{gas}$) dominates, and at a relatively weak coupling of M–A_{ads} the irreversible selection of A_{ads} should prevail, or in the atomic state ($A_{2gas} \rightarrow 2A_{ads} \rightarrow 2A_{gas}$), if $L_M > D_{MA}$, or in the form of compounds with metal atoms of the substrate ($A_{2gas} \rightarrow 2A_{ads} \rightarrow 2MA_v$), if $L_M < D_{MA}$. Even if the explicit energy preference – the thermal desorption mechanisms comprising a step of the pair recombination – are implemented only when at their lifetime adparticles have enough time to overcome the distance separating them, that is, they have a sufficient mobility and before they are desorbed experience multiple mutual "collisions." The study of the surface diffusion using electron and ion projectors, as well as model calculations, show that the activation barrier, which must be overcome by diffusing particles to move from one center to the adjacent one, usually does not exceed 10–30% of the value of its binding energy Q. In other words, the potential relief of metal substrates is quite "smooth," so that even particles strongly chemisorbed on the surface rapidly migrate at relatively low temperatures [9, 16]. If the adfilm is rarefied and adparticles are randomly distributed among nodes of equivalent adsorption sites, forming an ideal lattice gas, the recombination rate is proportional to the concentration of recombining particles and can be written as:

$$J_R = k_R \theta_A \theta_B . \tag{14}$$

Here, $k_R = k_{R0}\exp(-E_R/RT)$, and preexponential factor is typically rated as $k_{R0} \simeq (kT/n)n_m \approx 10^{28}$ at $cm^{-2}\ s^{-1}$ [9, 13]. The resulting molecules, AB_{ads}, depending on the relative strength of the adsorption and intermolecular bonds can then, with a certain probability, dissociate again or can be desorbed into the gas phase. Therefore the total kinetic scheme of desorption of the dissociated phase ($A_{ads} + B_{ads}$) is as follows:

$$A_{ads} + B_{ads} \rightleftharpoons AB_{ads} \rightarrow AB_{gas} \tag{15}$$

The rate of dissociation and desorption of molecules AB_{ads} as first-order reactions, are given by:

$$J^D = k_D P_{00} \theta_{AB}, J^d = k_d Q_{AB} \tag{16}$$

Here, $k_D = k_{D0}\exp(-E_D/RT)$, $k_d = k_{d0}\exp(-E_d/RT)$, and $P_{00} \simeq (1 - \theta_A - \theta_B)^2$ is the probability of finding two unoccupied centers in the vicinity of the molecule AB_{ads}. In the steady state, when the fluxes of recombination J^R, dissociation J^D and desorption J^d are mutually balanced, that is, in the quasi-stationary approximation $J^R - J^D \simeq J^d$, then the total flux density of desorption of molecules AB is given by:

$$J^d = \Gamma^\Theta \Theta_A \Theta_B, \tag{17}$$

where $\Gamma_\theta = k_R k_d/(k_d + k_D) = \{k_R,\ k_d \gg k_D;\ k_R k_d/k_D,\ k_d \ll k_D\}$, the effective rate constant of desorption of AB from the rarefied ($P_{00} \rightarrow 1$) dissociated phase ($A_{ads} + B_{ads}$). In the limiting case $k_d \gg k_D$, the surface reaction $A_{ads} + B_{ads} \rightarrow AB_{ads}$ is practically irreversible, so that the process of the separation of molecules AB from the dissociated phase is controlled by the rate of recombination of A–B and is characterized by the effective activation energy of $E_d' \simeq ER \geq Q_{AB}$. In the opposite limit case $k_d \ll k_D$, the quasi-equilibrium $A_{ads} + B_{ads} \rightleftharpoons AB_{ads}$ is set in the adfilm. Against this quasi-equilibrium, the separation process of AB is controlled by the rate of desorption from the molecular state $A_{ads} \rightarrow AB_{gas}$ and characterized by the effective activation energy $E'_d = E_R - E_D + E_d = Q_{AB}$.

2.4. Thermal desorption of the metal–hydrogen and metal–nitrogen systems

Hydrogen and nitrogen, chemisorbed on the surface of the transition metals in the atomic state (dissociative), almost always are present in the gas phase in the molecular form (reversibly). The resulting pair recombination of adatoms H_{ads} or N_{ads} and the subsequent desorption of the weakly bound with the metal substrate molecules of H_2 and N_2 at kinetic reactions of the second order [13] is:

$$H_{ads} + H_{ads} \rightarrow H_{2gas} \tag{18}$$

$$N_{ads} + N_{ads} \rightarrow H_{2gas} \tag{19}$$

$$J_D = k^\theta \theta^2 \tag{20}$$

This is consistent with the energy criterion (12), according to which the dissociating diatomic adsorbate desorption occurs predominantly in the atomic state, if the binding energy of the adatom and the metal substrate Q exceeds the dissociation energy of the molecules in the gas phase D ($Q > D$) and in the molecular state in the case $Q < D$. Table 2 shows the binding energy of adatoms and the heat of chemisorption of hydrogen and nitrogen on an atomically clean surface of some transition metals. It is seen that values of Q_H and Q_N growing among Pt, Ir, Pd, Ni, Fe ≤ Re <Mo ≤ W <Nb <Ta satisfy the inequalities:

$$Q_{H2} < Q_H < D_{H2} \tag{21}$$

$$Q_{N2} < Q_N < D_{N2} \tag{22}$$

Therefore, desorption of nitrogen and hydrogen in a molecular state is energetically advantageous process.

Metal	Hydrogen, D_{H2} = 103 kcal mol^{-1}		Nitrogen, D_{N2} = 225 kcal mol^{-1}	
	Q_{H2},kcal mol^{-1}	Q_H, kcal g-at^{-1}	Q_{N2}, kcal mol^{-1}	Q_N, kcal g-at^{-1}
Tantalum	50	77	130-140	130
Niobium	-	-	120-130	175
Tungsten	33	68	80-85	155
Molybdenum	30	65	80-85	155
Rhenium	-	-	60-70	145
Iron, Nickel, Palladium, Iridium, Platinum	10-30	50-65	35-55	130-140

Table 2. The binding energy and heat chemisorbtion of hydrogen and nitrogen on atomically clean surface of transition metals

2.5. Thermal desorption of the metal–oxygen systems

In contrast to the reversible hydrogen and nitrogen, the evolution of molecular oxygen ($2O_{ads} \rightarrow O_{2\ ads} \rightarrow O_{2\ gas}$) is observed only in some precious metals, which is characterized by a relatively low binding energy of the oxygen adatoms with the metal substrate $Q_O < D_{O2}$ = 118 kcal mol^{-1} [17]. On the most transition and, in particular, refractory metals oxygen chemisorbs with relatively high binding energies $Q_O > D_{O2}$ (Table 3).

Metals	*Q_{O2}, kcal-mol*	*Q_O, kcal g-at^{-1}*	*Q_O, kcal g-at^{-1}*
	Calorimetry		Mass-spectrometry
Titanium	238	178	-
Tantalum	214	166	179
Niobium	210	164	168
Tungsten	196	157	140
Molybdenum	173	146	-
Rhenium	-	-	127

Table 3. Binding energy of adatoms and heat of chemisorption of oxygen molecules

The basic mechanisms of the oxygen removal depending on the nature of the metal substrate are:

1. The direct desorption of oxygen adatoms in the atomic state $O_{ads} \rightarrow O_{gas}$.

2. The formation and desorption of metal oxides of the different stoichiometry $M+nO_{ads} \rightarrow MO_{n\ gas}$.

Data on the composition of the particles desorbed from the surface of refractory metals of Groups 4–7 after the oxygen chemisorption [14] are shown in Table 4. At low degrees of filling, oxygen desorbs, or in the atomic state, or in the form of monoxide MO. Wherein the first type (I) of desorption ($O_{ads} \rightarrow O_{gas}$) is observed in metals of Groups 6 and 7 (molybdenum, tungsten, rhenium) with the low-volume oxygen solubility, the second type (II) of desorption ($M + O_{ads} \rightarrow MO_{gas}$) in metals of Groups 4 and 5 (zirconium, vanadium, niobium, tantalum) is characterized by high-volume oxygen solubility. In both the cases, the high-temperature desorption of oxygen from the rarefied adsorbed film proceeds as the first-order reaction:

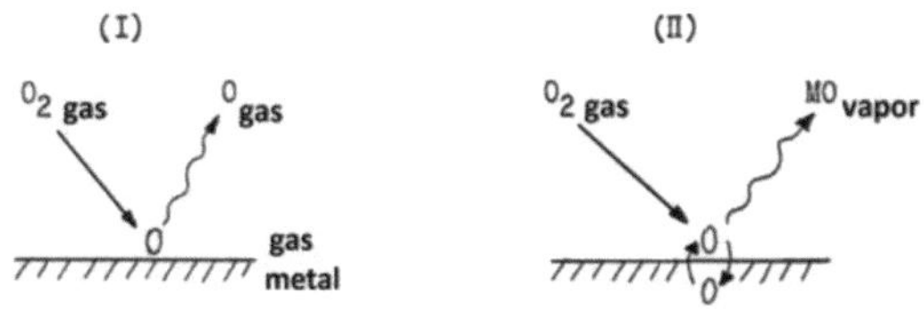

With an increasing degree of filling, the contribution of flows of polyoxygen molecules MO_A and MO_3 rapidly grow, and become predominant in the monolayer coatings $\theta \approx 1$. According to a generalized model of the interaction of oxygen with the transition metals [14, 15], and later confirmed by experiments for the systems of tungsten–oxygen, rhenium–oxygen, molybdenum–oxygen, niobium–oxygen, and tantalum–oxygen, that the formation of oxides proceeds in several successive stages, each of which is in the quasi-equilibrium. A limiting stage of the transition of oxygen into the gas phase consisting of oxides MO, MO_2 and MO_5 is their desorption:

$$O_{ads} + M \rightleftharpoons MO_{ads} \rightarrow MO_{gas}$$
$$O_{ads} + MO_{ads} \rightleftharpoons MO_{2ads} \rightarrow MO_{2gas}$$
$$O_{ads} + MO_{2ads} \rightleftharpoons MO_{3ads} \rightarrow MO_{3gas}$$

The effective activation energy $E'_{d,\mathrm{MO}}$ of oxygen releasing from the atomic phase in the composition as oxides MO_n is expressed by

$$E'_{d,\mathrm{MO}n} = nQ_0 - L_{\mathrm{M}} + D_{\mathrm{MO}n}$$

where L_{M} is the heat of the metal sublimation, $D_{\mathrm{MO}n}$ - MO_n the oxide dissociation energy in the gas phase ($MO_n \rightarrow M_{gas} + nO_{gas}$). Depending on the sign of the difference $D_{\mathrm{MO}} - L_{\mathrm{M}}$ (the values of $D_{\mathrm{MO}} - L_{\mathrm{M}}$ are given in Table 4), the value $E'_{d,\mathrm{MO}}$ is more or less Q_0. Consistent with the simple energy criteria set out above, rhenium, molybdenum, and tungsten emit predominantly the atomic oxygen characterized by values $L_{\mathrm{M}} > D_{\mathrm{MO}}$, whereas titanium, zirconium, vanadium, niobium, and tantalum emit predominantly monoxides characterized by $L_{\mathrm{M}} < D_{\mathrm{MO}}$

On the basis of data on the oxygen solubility and the oxides desorption during the vacuum metal processing, the possibility of the oxygen removal (deoxidation) is evaluated by an evaporation ratio that is meaningful of the relation of the saturated vapor pressures of the

Metal	Desorbing particles		Temperature of the desorption start, K	D_{MO} - L_M, Kcal mol^{-1}	Maximal solubility of oxygen in metal, at.%
	θ«1	θ≈1			
Titanium	TiO	TiO_2	-	44	32
Zirconium	ZrO	ZrO_2	>1900	36	30
Vanadium	VO	-	>1800	25	15
Niobium	NbO	NbO_2	>1900	7	11
Tantalum	TaO	TaO_2	>2100	7	8
Molybdenum	O(Mo)	MoO_2, MoO_3	>1700	-41	0.15
Tungsten	O	WO_2, WO_3	>1700	-46	0.06
Rhenium	O	(ReO_2) ReO_3	>1300	-47	0.01

Table 4. Composition of desorbing particles from the surface of refractory metals [14]

gaseous oxides and the metal over the metal surface in a vacuum, for the temperature interval 1800–2000 K (for which there are experimentally validated thermodynamic data) [18]. A serious disadvantage of this assessment is that it does not take into account the possibility of the high atomic oxygen desorption from molybdenum and tungsten. In any case, the currently available information concerning the thermal oxygen desorption from refractory metals reveals the following: Group 4 metals (titanium, zirconium, hafnium) have the greater oxygen solubility, the increased affinity for oxygen and the desorption of oxygen from these metals is in the form of oxides (monoxides); wherein the evaporation ratio is about 2–3, and the possibility of the oxygen removal from these metals due to desorption is extremely small. Titanium at $T > T_m$ is also poor due to evaporative deoxidation, but in the case of melting zirconium and hafnium it is possible to achieve a certain purification of oxygen. From Group 5 of metals (vanadium, niobium, tantalum), oxygen desorbs in the form of monooxides (suboxides) and the evaporation ratio is 28–42, that is, purification of these metals from oxygen by the monoxide desorption due to high-temperature annealing in vacuum is possible. The possibility of evaporative deoxidation in a more simplified form was shown by Ono and Moriyama [18] and Kulikov [19]. All common industrial metals were divided into three groups according to their capacity for "self-deoxidation" (Table 5). Note that in those studies the possibility of removing oxygen from molybdenum and tungsten in the atomic state has not been considered.

In the assessment [19] carried out for molten niobium and tantalum in vacuum, the evaporation ratio reaches 2,000–3,000, indicating intensive removal of oxygen from liquid metals of Group 5 in vacuum. In this case, the potential of the deep refining of tantalum at $T > T_m$ is much higher than that of niobium. An exception in this group is the system of vanadium–oxygen, which prevails in the vapor pressure of vanadium and, respectively, the evaporation rate is only about 4. By the thermodynamic assessment, the vapor pressure of gaseous vanadium oxides VO and VO_2 at $T > T_m$ is well below the saturated vapor pressure of vanadium over the melt. Therefore, evaporative deoxidation of liquid vanadium unlikely. The situation is similar to the case of

http://dx.doi.org/10.5772/59397

Evaporative deoxidation is possible	Evaporative deoxidation is impossible
$MoO/Mo = 10^{0.5}$	$Ti/TiO = 10$
$NbO/Nb = 10$	$V/VO = 10^2$
$BO/B = 10^2$	$Be/BeO = 10^3$
$WO/W = 10^2$	$Cr/CrO = 10^4$
$ZrO/Zr = 10^2$	$Mn/MnO = 10^5$
$ThO/Th = 10^3$	$Fe/FeO = 10^6$
$HfO/Hf = 10^4$	$Ni/NiO = 10^7$
$TaO/Ta = 10^4$	

Table 5. Capacity of refractory metals for evaporative deoxidation

chromium. However, tungsten and molybdenum, other members of Group 6, are characterized by the small oxygen solubility, by the low activity of dissolved oxygen, and by extremely high coefficients of evaporation, indicating a strong tendency of these metals to oxygen removal by the evaporation of oxides (oxygen). The major oxides are desorbed molecules like MO_3 and their dimers. At low content of oxygen in liquid molybdenum and tungsten at $T > T_m$ the oxygen should generally be present above the melt in the atomic state and the gaseous oxides desorption is negligible. The thermodynamic evaluation showed that at the removal of the diffusion supply of oxygen from the bulk of the liquid metal to the surface, from which desorption occurs, the deep removal of oxygen can apparently be achieved as a result of the vacuum treatment of liquid molybdenum and tungsten. The simple qualitative assessment of the high-temperature behavior of the metal–oxygen system in vacuum (and accordingly, the high-temperature metal processing, such as annealing and exposure in the molten state for the purpose of oxygen removal) can be divided into three types. In the first type (titanium, vanadium, chromium) oxygen desorption (or oxides) does not occur because the oxygen is strongly associated with the metal lattice and has the high thermo-chemical stability, and the metal itself has the vapor pressure greater than the vapor pressures of any oxides formed in this system. In the second type (molybdenum, tungsten), the vapor pressure of the oxides is much higher than the metal vapor pressure. In this case, the oxygen chemisorbed is easily desorbed alone or reacts rapidly with the metal and forms oxides, which are desorbed. The oxygen does not even dissolve in the metal lattice. The third type (niobium, tantalum, zirconium, and hafnium) is an intermediate between the first two.

2.6. Thermal desorption of the metal–carbon–oxygen

We have already noted that carbon is very strongly retained on the surface of metals and, unlike nitrogen and hydrogen, cannot be desorbed "independently" even from the surface of molten metal. The main mechanism of carbon release during vacuum annealing is the formation and desorption of CO molecules as a result of pair recombination of carbon and oxygen adatoms by the reaction [17]

$$C_{ads} + O_{ads} \rightleftharpoons CO_{ads} \rightarrow CO_{gas}$$

At sufficiently high temperatures of the metal substrate and the low contents of adsorbed particles when the adatom phase can be considered [20] as a two-dimensional gas ($C_{ads} + O_{ads}$), the resulting flux density of the CO desorption is given by $J^d = \Gamma^{\theta}\theta_o\theta_c$, where $\Gamma^{\theta} = (k_R k_d / k_D)\sigma$ is the effective constant of the CO desorption rate from the dissociated phase ($C_{ads} + O_{ads}$) and $\sigma=(1+k_d/k_D)^{-1}$ the probability of dissociation of the chemisorbed molecules CO_{ads}. In the limiting case $k_d >> k_D$ ($\sigma \rightarrow 0$), practically all the formed molecules (CO_{ads}) are desorbed into the gas phase, so CO removal is controlled by the rate of recombination of the adatoms C_{ads} and O_{ads} and characterized by an effective activation energy $E'_d \simeq E_R$. In the opposite limit $k_d \ll k_D$ ($\sigma \rightarrow 1$), the quasi-equilibrium $C_{ads} + O_{ads} \rightleftharpoons CO_{gas}$ is established in the adfilm, so the process is controlled by the release rate of the CO desorption from the molecular state and is characterized by the effective activation energy $E'_d \simeq E_R + E_d - E_D$. The question is which of these mechanisms is realized on the surface of the transition metals. It is closely linked to the question of the strength and character of the CO bond with the metal substrate [9, 21]. It is found that unlike molecular hydrogen, nitrogen and oxygen, the molecules of CO are firmly chemisorbed on the surface of transition metals to form the so-called β-structure state with one of the following types: (a) the bridge structure 2M-CO, and the CO molecular axis is perpendicular to the metal surface, and (b) the structure of the two-point adsorption of M-C-O-M, wherein the axis of the CO molecule is parallel to the metal surface.

Already in the early studies, it is found that, despite the obvious weakening of the intramolecular bond of C–O, the β-chemisorbed molecules of CO on many transition metals have a high stability and do not dissociate substantially during the chemisorption process. For a long time this was considered to be the molecular nature of the chemisorption of CO on transition metals, as opposed to the distinct dissociative chemisorption of hydrogen, nitrogen, and oxygen. Recently it has been shown [22] that, depending on the nature of the metal and the heat of chemisorption Q_{CO}, the β-chemisorption of CO can be of both molecular and dissociative character clearly. It is assumed that there exists a critical value of the chemisorption heat ($Q_{CO} \geq 65$ kcal mol^{-1}), from which the CO molecules show a tendency to dissociation. Table 6 presents the calorimetric data on the initial heats of the CO chemisorption on films of certain transition metals [21] specifying the nature of the chemisorption observed in each case.

The CO chemisorption on transition metals (platinum, palladium, rhodium, iron, nickel, cobalt, etc.) of Group 8 and rhenium of Group 7 is relatively weak ($Q_{CO} \leq 60$ kcal mol^{-1}). So, the CO molecules on the surface of these metals are sufficiently stable and hardly dissociate due

Metal	Q_{CO}, kcal mol⁻¹	Chemisorption mode
Titanium	155	Dissociative chemisorption with dissolving in metal lattice
Zirconium	150	
Niobium	123	
Tantalum*	110	
Tungsten	85-90	Dissociative chemisorption (?)
Molybdenum	70-80	
Rhenium*	62	Molecular chemisorption
Platinum	50	
Cobalt	48	
Rhodium	47	
Iron	45	
Palladium	43	
Nickel	42	

*Q_{CO} for Tantalum, Rhenium were received by flashing

Table 6. Initial heat and character of chemisorbtion of CO on some transition metals [21]

to moderate temperatures and the presence of high energy barrier to dissociation of CO, the value of which exceeds the binding energy of CO–metal [16, 22]. When the metal substrate is rapidly heated to the high temperature (flashing), the entire β-CO layer chemisorbed at room temperature has enough time to be desorbed before the dissociation rate becomes significant. The thermodesorption process obeys first order kinetics, i.e., it is molecular [9, 21, 23]. Refractory metals of Group 6 (tungsten and molybdenum) chemosorb CO quite strongly ($Q_{CO} \simeq 70 \div 90$ kcal mol^{-1}), so the dissociation probability of the CO molecules on the surface of these metals, apparently, is significantly higher than in the previous case, although the nature of the CO chemisorption on tungsten and molybdenum is still debatable. Numerous data obtained on tungsten and molybdenum using a flash traditionally are interpreted as an evidence of the molecular nature of the β-CO chemisorption (exceptions include [24, 25], where the dissociative model is discussed). On the other hand, Gillet et al. [20] and Felter and Estrup [26] studied chemisorptions energy by electron diffraction using Auger electron spectroscopy, photoelectron spectroscopy, and energy losses. Thus the results of these studies present a strong evidence for the dissociation of the CO molecules in β-chemisorption on molybdenum and tungsten. Summarizing the results available, Guillot et al. [27] concluded that at the flash the β-phase likely goes into the dissociated state and without dissolving in the metal lattice, "flies" off from the surface as a result of the recombination desorption. Refractory metals of Groups 5 and 6 (titanium, zirconium, vanadium, niobium, tantalum), dissolving interstitial impurities well, chemisorbed CO so firmly ($Q_{CO} \geq 100$ kcal mol^{-1}) that already at room temperature, the rapid dissociation of the CO molecules is observed on the surface of these metals. The dissociation flows substantially by inactivated adatoms and accompanied by dissolving oxygen and carbon in the volume of the metal substrate [28-32]. During a flash, long before the intense CO desorption, in the adfilm and between the adsorbed particles and the particles dissolved in the surface layer, a quasi-equilibrium is established: $C_{ads} + O_{ads} \rightleftharpoons \beta\text{-}CO$; $C_{ads} \rightleftharpoons C_{dis}$; $O_{ads} \rightleftharpoons O_{dis}$, against which the dissolution and diffusion of carbon and oxygen into the sample occur. The subsequent CO desorption is of a "two-phase" character: after the removal of the thermal dissociated surface phase (a fast step) the CO removal is observed as a result of diffusion of solute atoms of carbon and oxygen on the surface, which appears in the spectra of the thermal desorption as slowly decaying "diffusion tails" [31].

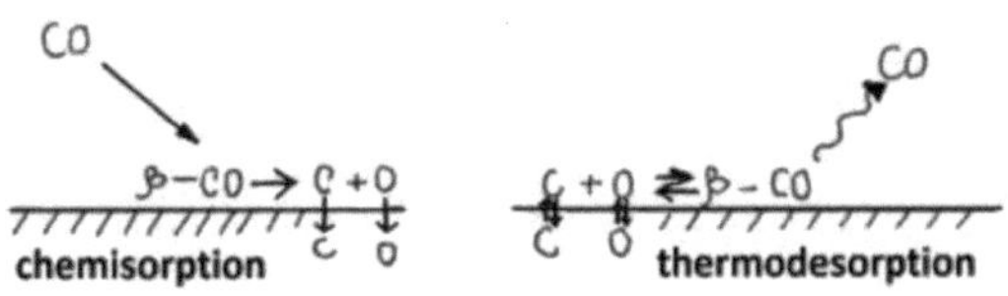

Thus, the results of studying the interaction of CO with transition metals indicate that the reaction $C_{ads} + O_{ads} \rightarrow CO_{ads}$ on the surface of platinum, palladium, rhodium, nickel, iron, cobalt, and rhenium proceeds virtually irreversible ($k_d >> k_D$). Therefore, the controlling step is the release of CO from the dissociated phase ($C_{ads} + O_{ads}$). On the surface of such refractory metals as vanadium, niobium, tantalum, titanium, and zirconium, the quasi-equilibrium $C_{ads} + O_{ads} \rightleftharpoons CO_{ads}$ has been set in the adfilm against which the CO release in the gas phase is controlled by the desorption rate from the molecular state $CO_{ads} \rightarrow CO_{gas}$ ($k_d << k_D$). Apparently likewise the CO release proceeds from the surface of molybdenum and tungsten, although currently this view is not universally accepted.

2.7. Refining refractory metals from carbon and oxygen

A question on the reasons why some metals in the liquid and solid state cannot reach the thermodynamically predicted traces of carbon and oxygen during vacuum processing is repeatedly discussed in the literature. The studies of the interaction of liquid iron, niobium, and molybdenum with gaseous CO are performed, and attempts were made to eliminate the main drawback inherent in earlier works – the contact of the melt with the material of the crucible. Unfortunately, all of the experiments and their interpretations are clearly insufficient for an unambiguous and consistent description of the processes occurring in these complex systems, and have another drawback: a holistic approach to metals of the "iron triad" and to the refractory metals of Groups 5 and 6. All these metals are fundamentally different objects in terms of the removal mechanisms of carbon and oxygen. Iron, cobalt, and nickel have a single, common for carbon and oxygen "channel" in the form of the CO (CO_2) desorption. From refractory metals, oxygen can be desorbed together with carbon and in the atomic state, as well in the form of oxides. Thus, if the appearance of the "limit" contents during degassing of iron and its analogues is difficult to explain within traditional concepts of the reaction, at the degassing of refractory metals, carbon content should be limited is an obvious necessity. On the other hand, since carbon "alone" substantially cannot be desorbed from refractory metals

http://dx.doi.org/10.5772/59397

at high temperatures, the carbon release during vacuum processing occurs only as a result of its reaction with oxygen diffusing to the metal surface.

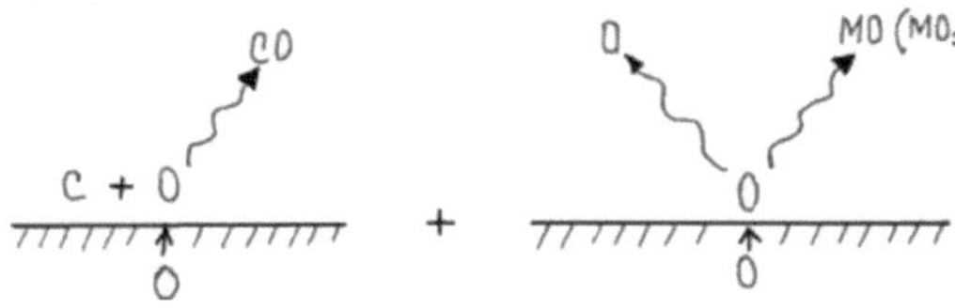

The question of what are the patterns of this interaction, which is one of the central problems of obtaining high-purity refractory metals, is repeatedly discussed. However, in contrast to the above two-component solutions of metals with hydrogen, oxygen, or nitrogen, the consistent theoretical analysis of the kinetic behavior of carbon and oxygen in the thermal desorption from solutions M-O-C hitherto substantially is not performed. Because of this, a reliable basis for making and interpreting relevant experiments is missing, and the main reference for the forecasting decarbonization capabilities of refractory metals in vacuum is the study of a thermodynamic equilibrium of the reaction $C + O \rightleftharpoons CO_{gas}$. Table 7 summarizes the equilibrium constants of this reaction on thin films of tantalum and niobium and on levitated liquid droplets of niobium and molybdenum [33–37]. These data show that the equilibrium pressure P_{CO} over solutions of (C+O) in tantalum, niobium, and, especially, molybdenum is very high (Figure 2). Thus, it is supposed [34, 37–41] that when an excess of oxygen (compared to carbon) is added to the solution, the vacuum processing of those metals should be finished with very low contents of carbon and oxygen.

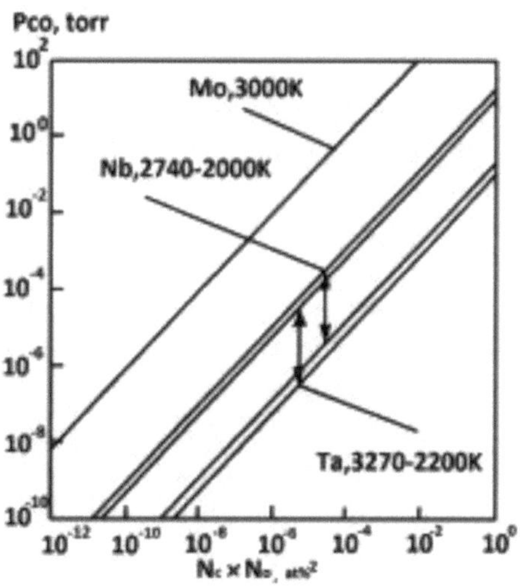

Figure 2. The equilibrium pressure of CO above the carbon and oxygen solutions in niobium, tantalum and molybdenum

According to Figure 2, at a pressure of 10^{-8} Torr the equilibrium contents of the CxO product in the solution for niobium varies from ~7 × 10 (at.%)^2 at 2000 K to ~9 × 10 (at.%)^2 at $T \approx T_m$ and for molybdenum, the value is ~(10–12) × 10 (at.%)^2 (at.%)^2 at 3000 K. Meanwhile, attempts to establish a certain correlation between the ratio O/C of the concentrations of carbon and oxygen in initial samples, and a minimum concentration of carbon, which can be achieved under UHV annealing or melting niobium, vanadium, and molybdenum, have shown that even the sample contains the considerable supersaturation with oxygen (O/C ≥ 10), the carbon content in the volume cannot be reduced more than 10–10^2 times, and to less than 10^{-2}–10^{-3} at.%. In those cases where the initial carbon content is ~10 at.% and lower, the introduction of oxygen into the solution up to 0.5 at.% does not lead to any appreciable reduction of the carbon content in general, although P_{CO}, corresponding to the current product of the carbon and oxygen contents in the solution, is higher on several orders of magnitude than the actual vacuum in the vacuum chamber. To explain these results, the suggestion is made that after achieving certain "super-equilibrium" product of contents $(CxO)_{crit}$, the surface reaction $C_{ads} + O_{ads} \rightleftharpoons CO_{gas}$ is slowing down sharply [36].

System	Constant of reaction, K_p, *(at.%)^2/torr*	Temperature, K
Nb-C-O	6.4 × 10^{-7}exp(67 kcal (RT)$^{-1}$	1,900–2,200
	0.21	1,850
Ta-C-O	8.3 × 10^{-6}exp(65 kcal (RT)$^{-1}$	2,000–2,300
Mo-C-O	1.5 × 10^{-4}	3,000

Table 7. Equilibrium constants of reaction C + O $\rightleftharpoons$ CO for some refractory metals

The concepts are developed according to which such situation would occur when the characteristic time between acts of the recombination of adatoms of carbon and oxygen becomes equal to the characteristic lifetime of some hypothetical active sites necessary for the exchange of the excess energy of the excited CO molecules [38-41]. Without dwelling on the detailed critique of such ideas, we merely point out that they are clearly contrary to well-established provisions of the chemical kinetics, and attempts to analyze them on the basis of the facts have an openly fitting character. Some researchers have a different point of view and linked the "discovery" of a sharp slowdown of the interaction of carbon and oxygen at low concentrations of carbon to an insufficient sensitivity of conventional methods of analysis, located at the level of 10^{-2}–10^{-3} at.% of carbon. They convincingly demonstrate this by applying the activation method by ^{3}He ions for the registration of residual concentrations of carbon with a sensitivity of 10^{-6} at.%. Indirectly, the same data show a high ratio of the electrical resistivities at room and liquid helium temperatures, which could be obtained as a result of high vacuum annealing pre-oxidized foils or thin wires of niobium and tantalum. Kulikov [19], emphasized the influence of the kinetic factors on the decarbonization processes, and the lack of reliable experimental data on the interaction of carbon and oxygen in metals of Groups 5 and 6 are associated primarily with an insufficient regard of the following two factors: the specific

http://dx.doi.org/10.5772/59397

surface of the molten metal and the diffusion inhibition in its volume. The reduced specific surface of the melt leads to a proportional decrease in a rate of the refining and an increase in duration of the removal of oxygen and carbon. An increase in the diffusion paths affects the reduction of the diffusion flow of impurities to the surface desorption in a quadratic dependence. Describing the issue of the laws of interaction of carbon and oxygen in refractory metals in general, we can state the following:

a. The only reliable data that would reveal the intensity of the interaction of dissolved interstitials in refractory metals are obtained from the study of the thermodynamic equilibrium of the reaction $C_{dis} + O_{dis} \rightarrow CO_{gas}$ (the temperature dependence of the equilibrium constant, the heat of dissolution of CO in metals).

b. Except for a few attempts to empirically establish the correlation between the ratio of the initial contents of oxygen and carbon, and the residual carbon content in the solution, the interpretation of results is based primarily on an intuitive basis and is largely controversial work. The systematic experimental study of the kinetics of the interaction of carbon and oxygen in vacuum processing, which should be carried out, is almost absent. A clear picture of the overall behavior of oxygen and carbon is not substantially present.

This is due, on the one hand, to the lack of a consistent theoretical analysis of patterns of the thermal desorption of both interacting impurities of different kinds, that is, with the lack of a sound basis for setting and interpretation of the relevant experiments, and on the other hand, from a purely methodological difficulties of simultaneous recording of kinetic changes of the carbon and oxygen concentrations in the process of annealing or melting. Studies on obtaining pure refractory metals in vacuum by electron beam melting (EBM) and by electron-beam floating zone melting (EBFZM) have been done for a long time. Since EBM and EBFZM have some significant differences (EBM – the presence of the mold, the drip process; EBFZM – the float-zone melting, a convective mixing of the liquid zone), it seems reasonable to consider some results of the studies using these techniques separately.

2.8. Refining refractory metals by EBFZM

If the removal of nitrogen and hydrogen (and in some cases, of oxygen) from refractory metals is the ample high-vacuum treatment of solid or molten samples, for the removal of carbon the rationality of using this treatment remains problematic. Difficulties are encountered in obtaining the "carbon-free" refractory metals of Groups 5 and 6, and a lack of clear understanding of the nature of the processes occurring in their decarburization lead to the fact that the problem of the removal of carbon (as, indeed, and oxygen) is seen as "the cornerstone" of metallurgy of pure refractory metals. Although the issue of the removal of oxygen from these metals is more or less clear, however, the existing experimental data suggest rather an inability to remove oxygen from the metals of Groups 5 and 6 by the evaporative deoxidization or the deoxidization by carbon. Apparently, there is a need to emphasize once again a very important aspect of the problem. Data on the kinetics and thermodynamics of these processes do not have an experimental evidence due to the limited opportunities of the existing melting methods and the absence of clear theoretical concepts, but also because of the low sensitivity of the analytical methods used. Essentially, none of the known conventional physical-chemical analysis methods (vacuum extraction, mass-spectrometry, the coulometric method, and so on) do not allow for a reliable analysis of carbon and oxygen at concentrations below ~10^{-3}%. Using the method for determining the relative residual resistance (RRR) only allows to characterize the purity of the metal integrally, but in any case do not provide insight into the behavior of the impurities content. Perhaps only the emergence of new nuclear-physical methods of analysis with the removal of the contaminated surface layer would clarify the details of the behavior of individual impurities in low concentrations.

Niobium. Experimental studies aimed at obtaining high-purity niobium showed a steady growth of RRR, while reducing the amount of interstitial impurities [42–44] and the volatility of metal impurities (tantalum, tungsten) [45]. In early stages of research for refining processes of niobium there were used only EBFZM and EBM methods. The contribution of various metallic and nonmetallic elements in the value of RRR is studied: it is shown that the most significant is the effect of interstitial impurities (Fig. 3):

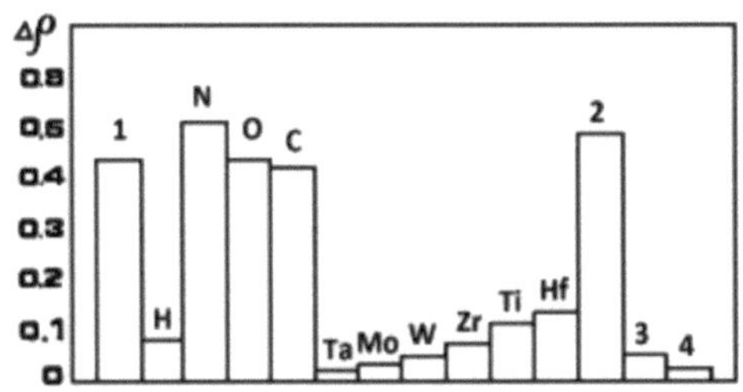

Figure 3. Influence of metal impurities and gas interstitials on RRR [42].

It is found that the effect of the zone processing by EBFZM of niobium plays a minor role (as, indeed, in the case of other refractory metals) [43]. During EBFZM, the molten metal zone is protected from the interaction with the components of the gas phase in the "cloud", formed of niobium vapors. While in accordance with the thermodynamic and kinetic data it is more advantageous for impurities to be removed from the liquid phase than from the solid; however, the lowest concentration of interstitial impurities are obtained by degassing niobium samples in the solid state. Furthermore, during vacuum melting zone the continuous interaction between solid portions of the heated sample with residual gases can occur, i.e., it is possible that interstitials would dissolve again in the metal at $T<T_m$. Many researchers used the high-temperature UHV treatment [42–45]; however, it turned out that it is impossible to obtain a high-purity niobium, even with such processing, without using other purification methods of niobium in parallel. As a result of the detailed analysis of the available experimental data, we propose the following scheme for the production of high niobium: the chemical purifying niobium powders, compacting of niobium powders, EBM of the PM rods in a high vacuum, electrolytic refining of vacuum-melted rods, EBFZM of the compacted feeds under high vacuum, and UHV annealing the cast rods (single crystals). However, even with such a

complex scheme, the "carbon-free" high-purity niobium could be obtained with a great difficulty, so that there is still no clear idea on the effective decarburization process of niobium. It is quite possible that by variation of the ratio C/O in the initial specimens of niobium and by variation of the partial pressure P_{CO} in the gas phase, the niobium specimens with a low content of both interstitials could be obtained. A number of studies determine the relationship between the equilibrium concentrations of oxygen in niobium and the oxygen partial pressure P_{O2} in the vacuum chamber. Even an attempt is made to evaluate the influence of the gas phase in a vacuum chamber at the carburization or decarburization of solid niobium specimens.

Molybdenum. The available experimental information on the influence of parameters of EBFZM such as the number of zone passes, the rate of recrystallization, the purity of feeds, vacuum, etc., on the behavior of non-metallic and metallic impurities in molybdenum is clearly insufficient. It is found that the predominant process of EBFZM of molybdenum is the evaporation of impurities while a minor effect is of the zone effect, confirming by the micro-hardness or RRR on the length of single crystals of molybdenum. Noted that after the multi EBFZM of molybdenum the single crystals have the high degree of purity: RRR values reach 15,000–30,000. However, there is a noticeable unevenness in the distribution of impurities along the length and radius of crystals: the single crystal of molybdenum of 120 mm long after eight zone passes at a rate of 5 mm/min has a value of RRR varied from 17,000 to 1,000. It is unlikely that this change of RRR occurs only due to the zone effect. Increasing the number of zone passes, and thus the exposure of molybdenum in the liquid state in a vacuum, results in an increased purity. However, results of many studies diverge markedly in the quantification of this effect. Greater certainty exists regarding the degree of vacuum: the purity of molybde-num significantly enhanced by the use of the oil-free vacuum and UHV.

Tungsten. The effect of interstitials on physical and mechanical properties of tungsten is significant. Therefore, repeated attempts to get a clearer picture of the behavior of these impurities (in particular, carbon and oxygen) during EBFZM are done. Usually, the first zone passage is the most effective in the purification of tungsten: the RRR value is at 21,000–25,000, and the oxygen content decreases from 2×10^{-3} to 3×10^{-4}%. Thus, the carbon content is reduced from 2.5×10^{-3} to $(4–20) \times 10^{-4}$%. Several studies have noted that multiple zone passes do not result in the complete removal of oxygen and carbon from tungsten (it seems this process is endless). However, these findings cannot be considered conclusive because the impurities are often found at the detection limit of the analytical methods (even of the most sensitive), and therefore could not give reliable information about the behavior of these impurities. It is shown that, depending on the quality of the vacuum and the growth rate of tungsten single crystals, the RRR values range from 13,600 to 99,300. An application of the activation by ^{3}He ions, allowing defining the content of carbon and oxygen to ~10^{-7}%, allow establishing a definite link of melting conditions with a final content of these impurities in tungsten. It turned out that the content of oxygen and carbon in the tungsten specimens does not exceed 1×10^{-4}at.%. The study of behavior of the same impurities by the same method in tungsten alloyed with rhenium shows that the carbon content in the tungsten after a zone refining does not dependent on the content of rhenium and is at 1×10^{-5}%, while the oxygen content is strongly dependent on the content of rhenium and is at a minimal level (~5×10^{-6}at.%) at 2–5% rhenium. At these contents of rhenium, the oxygen content is for about an order of magnitude lower than in pure tungsten melted under similar conditions. Some authors believe that there is an additional deoxidation by rhenium and this process is very similar to the deoxidized metals of the "iron triad" group by some element-reductants [19]. It should be noted that the majority of the studies using EBFZM is aimed mainly at obtaining and characterization of single-crystal samples, rather than on finding and understanding of a consistent behavior of interstitial impurities. Therefore, data on the behavior of interstitial impurities are few and uninformative, as the analytical methods used do not exclude the influence of the surface oxides and other contaminations of samples, which is very significant in the determination of small amounts of impurities, and therefore do not give a true picture. Therefore, it is necessary to conduct a correct and consistent study of the behavior of impurities (especially interstitial impurities oxygen and carbon) in metals such as niobium, molybdenum, and tungsten. Particular attention is paid to the systematic application of more sophisticated and sensitive analytical methods to remove surface contaminations from samples prior to analysis.

2.9. Theoretical model of the removal of interstitials from refractory metals

The results of our theoretical investigation of the interaction of carbon and oxygen in the solid and liquid refractory metals in vacuum are presented here briefly. Although the results are general in the nature, the presentation is mainly applicable to metals of Group 5 and 6 of the Periodic Table that are the most extensively studied in aspects of the interest to us. The basic model is given by sequence of the segregation of carbon and oxygen atoms from the volume of metals in a vacuum, as shown schematically in Figure 4 [46–53]. Earlier, during the discussion of the fundamental issues associated with the interaction of impurities a similar conceptual approach has been developed, using the similar notation, terms, and equations. This greatly simplifies the description and understanding of our theoretical model together with the fundamental issues made before us.

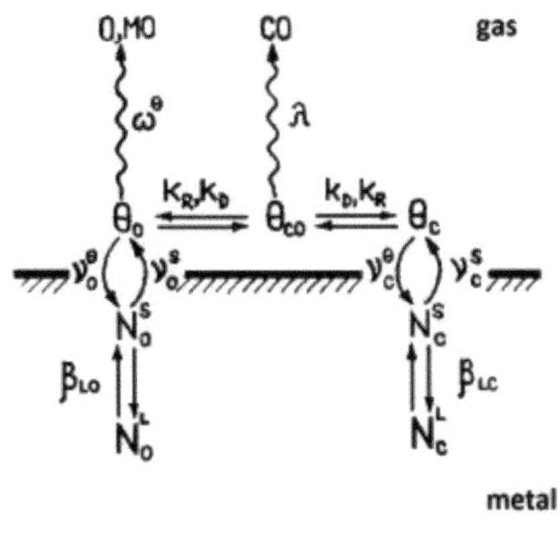

Figure 4. Model of the carbon and oxygen removal.

Carbon atoms diffuse into the subsurface layer and then bypass the phase boundary and on the surface recombine with the oxygen adatoms with the formation and subsequent desorption of the CO molecules. For oxygen adatoms impinging on the surface in a similar manner, there are two possibilities: either recombine with carbon adatoms and desorb as CO or forming a chemical bond with the metal atoms, and desorb as monoxide MO (at high concentrations of oxygen, in the form of MO_2). In formulating a system of nonlinear boundary value problems of diffusion, which allows to describe the kinetics of these processes, it is assumed that the metal that isothermally annealed in vacuum at a temperature T has the form of an elongated plate, of an extended half-thickness cylinder or a sphere of a radius R, and in the volume of metal in the initial moment of time the carbon and oxygen atoms with the contents of $N_c(0)$ and $N_o(0)$ are uniformly distributed. Boundary conditions for the corresponding one-dimensional diffusion equations are found by matching the flux densities of diffusion $J^D|_s$, segregation $J^{s\theta}$, and desorption J^d of each impurity on the metal surface (the principle of the quasi-equilibrium)

$$J_i^D\Big|_s = -\left(D_i / v_s\right)\left(\nabla N_i\right)\Big|_s = J_i^{s\theta} = J_i^d$$

are written with an assumption that (a) the adfilm is sparse, (b) the lateral interactions in the adfilm can be neglected and (c) the contents of both impurities in the solution are much lower than the solubility limit. The comparison of the flux densities of the carbon desorption $J_c^d = \Gamma^\theta \theta_c \theta_o$ and of the oxygen desorption $J_o^d = J_c^d + \omega^\theta \theta_c$ (where Γ^θ is the constant of the rate of the CO molecules desorption from the atomic phase $O_{ads} + C_{ads}$, ω^θ the constant of the rate of the monoxide MO desorption (for metals of Group 6, atomic oxygen), θ_i the degree of the surface coverage of the metal adatoms of grade i) shows that there is some critical degree of the surface coverage by carbon $\theta^* = \omega^\theta / \Gamma^\theta$, when the desorption of oxygen has been changed dramatically: in the overcritical region $\theta_c/\theta^* \gg 1$ oxygen desorbs mainly together with carbon, while in the subcritical region $\theta_c/\theta^* \ll 1$ oxygen desorbs predominantly "by itself" (as MO or O) and almost does not "feel" the presence of carbon in the metal. Thus, the value of θ^* sets the characteristic scale the covering of the surface, and the corresponding critical content of carbon $N^* = \omega/\Gamma$ is the characteristic scale of contents in the solution. Here, $\omega = \omega^\theta f_o$ and $\Gamma = \Gamma^\theta f_o f_c$ the effective constants of the desorption rate from the subsurface layer of metal in a vacuum, respectively molecules of MO (O) and CO, f_i the coefficient of surface segregation. With this type the oxygen content $\xi = N_o/N^*$ and the carbon content $\eta = N_c/N^*$ are introduced. Based on the assumption of the quasi-equilibrium of the surface segregation of oxygen and carbon, and after appropriate dimensionless procedures the boundary conditions are obtained, which have defined a system of nonlinear boundary value problems of diffusion of carbon and oxygen. Further study of this system is carried out in the following sequence: first, the subcritical region of the carbon contents is considered and focuses on the behavior of the average volume contents of carbon and oxygen, and the connection between them, the results are then summed for the whole content region of carbon and, finally, the disclosed features of behavior patterns of the surface content of carbon in the high temperature annealing in vacuum. A particular attention is given to the limit degree of the refractory metal decarburization $L = \eta(0)/\eta_\infty$, which can be achieved under specified conditions of the high vacuum processing (impurity composition, geometry and size of the sample, temperature, vacuum). It turned out that, contrary to traditional notions the decarburization degree L is determined by the oxygen supersaturation of the metal not on the initial carbon level, but on the characteristic for each metal a critical value $N^*(T)$. There is a strong dependence of the decarburization degree L from the effective size of the sample. Upon annealing of thin samples, the decarburization degree L is maximal and depends on $\xi(0)$ exponentially strongly and large even at the low oxygen saturations over the critical value N^*. Upon annealing of bulk samples, L is minimal and depends on $\xi(0)$ weakly root-like and is small even at the significant oxygen saturations over $N^*(T)$. The temperature dependence of the decarburization degree L is determined mainly by the temperature dependences of the critical concentration of $N^*(T)$ and the characteristic size of the sample $R^*(T)$. With decreasing the annealing temperature the value of N^* decreases sharply, and R^* increases dramatically: the first is equivalent to the "temperature" introduction of oxygen into the sample, and the second the "temperature" reduction of its effective size. As a result of the joint action of both factors, the decrease of the annealing temperature (within certain limits) should lead to a significant increase in the decarburization degree of the metal. An exposure to the gas phase becomes noticeable from the certain critical density of the adsorption flow α *. If $\alpha_{ads} \leq \alpha$ *, the annealing of the metal occurs in the same manner as in conditions of an absolute vacuum, that is, the carbon content always reaches its maximum level ($L^v \approx L$). In the case $\alpha_{ads} \leq \alpha$ *, it, before reaching the limit, passes through minimum ($L^v \approx L$), and then increases at a rate proportional to α_{ads}. In both cases, the high decarburization degree of L can be achieved only if $\alpha_{ads} << 1$, and at high flux densities $\alpha_{ads} >> \alpha$ the dependence of L on the initial supersaturation of the sample with oxygen dramatically weakens: in the limit of thin samples instead of the strong exponential dependence it becomes linear, and in the limit of massive samples it "disappears" at all. As the annealing temperature is lowered, the value of α_{ads} rapidly increases, and of α = decreases rapidly, so that below the certain temperature T_m the effect of adsorption becomes dominant and the decarburization degree L of metal falls. As increasing the oxygen content in the metal, as an increase in the effective pressure of CO in the system, leads to a shift in T_m to higher values. The quantitative results (in the form of appropriate tables and graphs) obtained in these studies provide the basis for selecting the optimum conditions for the high-temperature annealing of the refractory metals. Because the patterns of the carbon and oxygen interaction in the process of the thermal desorption of the liquid refractory metals are of a considerable interest for vacuum metallurgy, the issue is given special attention. A distinctive feature of the process of allocating the impurities of carbon and oxygen in the liquid metals, as compared with the solid ones, is complicated by a transport of them to the metal surface (molecular diffusion plus convective mass transfer). The intensity of this transport depends on the direction and rate of the hydrodynamic flow in the volume of the liquid metal. The basis for describing the kinetics of the process served a model of a δ-layer. Under this model, it is believed that mainly in the core melt due to convective mixing, the impurity is distributed uniformly with a certain concentration of N_L. The diffusion "resistance" or "braking" is mainly concentrated in a narrow subsurface layer of a thickness δ. Within this layer the transfer is carried out by the molecular diffusion with the flux density

http://dx.doi.org/10.5772/59397

$$J_{i\ s}^{D} = -\left(v_a \propto\right)^{-1} \dot{N}_L = \frac{\beta_L}{v_a}\left(N_L - \dot{N}_s\right) \tag{23}$$

Here N_s is the surface impurity concentration, α the specific surface of the melt, $\beta_L = D/S$ the mass transfer coefficient, which is the main parameter of the model and is generally determined experimentally. In typical cases, β_L is 10^{-1}–10^{-3} cm s^{-1} and δ is 10^{-2}–10^{-4} cm. A chain of equations (23) together with the consistency conditions of the flux densities of diffusion, segregation, and desorption of carbon and oxygen leads to the corresponding system of nonlinear differential equations, which shows that the previously defined patterns of the interaction between carbon and oxygen at the high-temperature annealing of refractory metals in the solid state in this study remain valid. In the case of the liquid metal, with the difference that in the case of the melt instead of the relative diffusion mobility of the impurities, the relative coefficient of the mass transfer occur and instead of the effective size of the solid specimen, the effective thickness of the δ-layer.

2.10. Experimental verification of the model

The experimental verification of the model was carried out on the solutions of carbon and oxygen in two metals (niobium, molybdenum) by two crucibleless melting methods: the high-frequency vacuum levitation melting (HF VLM) and the electron-beam floating zone melting (EBFZM). They are shown in Figures 5 and 6. In Figure 5 the liquid droplet (red) is shown inside the water-cooled cupper electromagnetic inductor (partially seen). The lines of the high-frequency electromagnetic field, which supports the levitated droplet, are shown by dashed lines. The volume of the levitated droplet is about 2.0–3.5 cm^{-3}, depending on the nature of the metal.

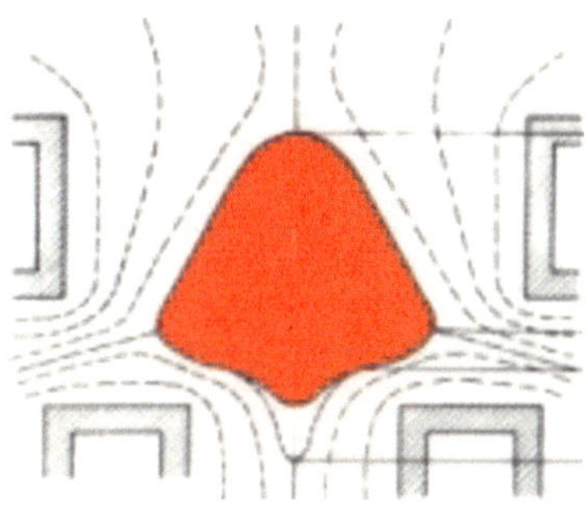

Figure 5. High-frequency vacuum levitation melting.

In Figure 6 the set-up for electron-beam floating zone melting is shown. The liquid zone is shown by red color. Here, the liquid zone does not contact with any contaminating materials like refractory crucible materials, except two interfaces with a feed and a grown crystal. The volume of the liquid zone is about ~2.0 cm^{-3} depending on the nature of the metal, dimensions of a growing single crystal, and the height of the liquid zone.

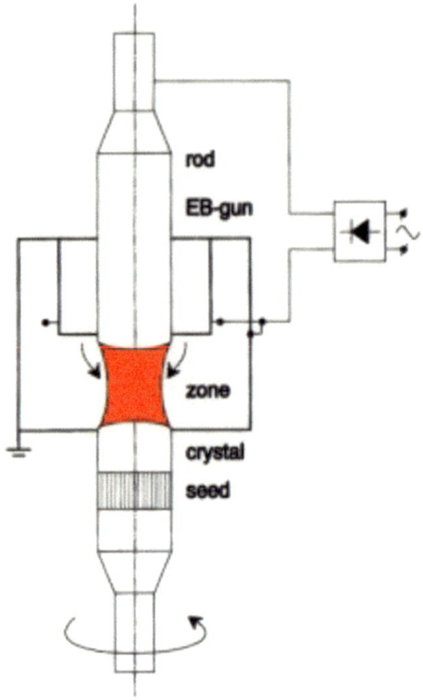

Figure 6. Electron-beam floating zone melting.

Specifically, the demonstration of validity of the basic provisions of the proposed model has been performed on niobium and molybdenum [53–55]. Both metals are excellent examples of the low solubility of carbon and oxygen, and the possibility of removing carbon by oxygen in an atomic state. In the case of niobium four series of samples with the oxygen content of 0.93 at.% and and the carbon contents of 0.35, 0.58, 0.83, and 1.05 at.% were prepared. The temperature of the levitated liquid drop of niobium is 2850 ± 50 K. The kinetics of the mean concentrations of oxygen and carbon in niobium for all four series are shown in Figure 7. It can be seen that the transition from the series 1 (O/C ≈ 9) to the series of 4 (O/C ≈ 6) leads to a sharp change in the slope of the curves in both graphs.

This, on the one hand, confirms the assumption of the high mass transfer rates in the levitated niobium drops and, on the other hand, indicates the occurrence of an experiment in a supercritical range of both impurities. In accordance with the model representations [46], an experimental verification of the model and a determination of the critical concentration N* is possible by simultaneous recording the behavior of the average concentrations of carbon and oxygen in the high-temperature vacuum treatment.

$$\partial \ln < N_c > / \partial \Delta \tag{24}$$

http://dx.doi.org/10.5772/59397

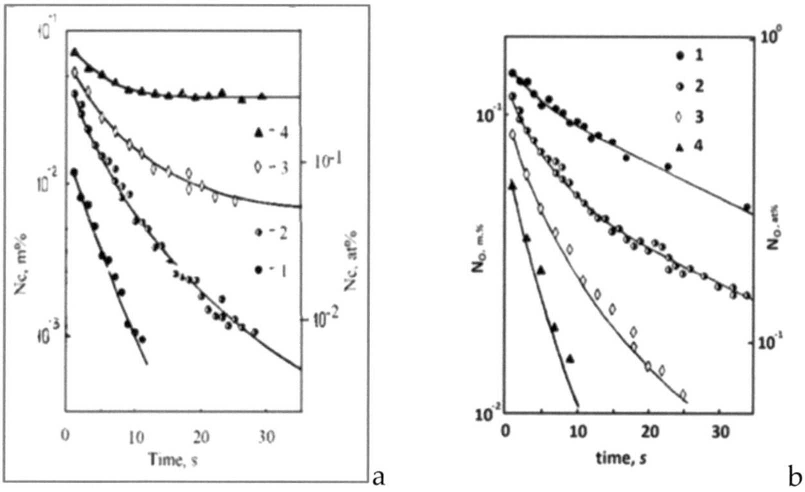

Figure 7. Kinetics of mean concentrations of carbon (a) and oxygen (b) in niobium.

Here, $\Delta = \langle N_o \rangle - \langle N_c \rangle + N_c^{\infty}$, and S is the parameter characterizing the relative intensity of diffusion and desorption flows. From expression (24) it is understood that the dependences $\langle N_c(t) \rangle$ and $\Delta(t)$, obtained by annealing of samples with different ratios of the initial contents of carbon to oxygen should "stack" on one line in the coordinates $\partial\Delta/\partial \ln\langle N_c \rangle - \Delta(t)$, with a slope $1/S$ and the initial ordinate N^*. In full accordance with the predictions of the model, the dependences $\ln N_c = f(\Delta)$, constructed from the experimental data (Figure 8), are close to linear and are almost parallel. The critical content, determined from the slope of the curves, taking into account the contribution of desorption of molecules NbO_2, is equal to $N^* = 0.05 \pm 0.01$ at. % and $S > 50$. Based on the values of N^* and S it is possible to directly compare the calculated and experimental carbon and oxygen behavior in time. Since when $S > 50$ the kinetics is "insensitive" to the value of the effective thickness of δ-layer, it is assumed that $q \to 0$ as $S \to \infty$ and the corresponding system of nonlinear differential equations are numerically solved, with $N^* = 0.05$ at. % and $t_0 = 43$ s (the relaxation time of the release of oxygen is found from experiments with samples of niobium doped with oxygen only). The initial conditions are used in the calculation of the content of oxygen and carbon in 1 second after melting, and then the results of numerical calculations are compared with experimental data. For all four series studied of the niobium alloy doped with both impurities there is a good agreement between theory and experiment. This is a direct experimental proof of the validity of the model in a wide range of impurity concentrations.

Experimental study of the removal of carbon and oxygen of niobium using sensitive methods of activation analysis (neutron and deuteron) are conducted in two directions: the study of the behavior of oxygen in carbon-free niobium and study of the joint behavior of carbon and oxygen in the process of multiple EBFZM. It has been established that oxygen and carbon are uniformly distributed over the entire length of the crystals except unsteady region in the initial portion of the crystal. The dependence of the full residual pressure in the melting chamber is

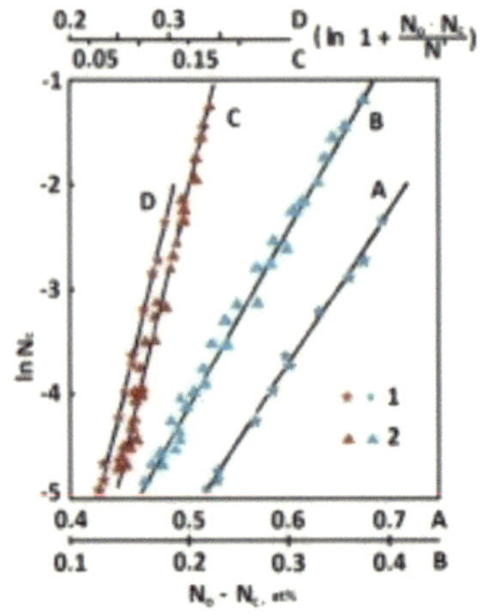

Figure 8. Dependences $\ln N_c = f(\Delta$, constructed on experimental data. In accordance with theory, they should be linear and parallel (A – B & C – D)

found from the number of passes and the warm-up mode of the melting chamber. Figure 9 shows the dependence of the oxygen content on the stationary portion of niobium single crystals of the number of zone passes for three preheat (I- 1 h, II- 3 hours, III- 5 h), obtained from the study of 23 crystals. At first, to illustrate the effect of surface contamination in the determination of small amounts of interstitials, the results of the determination of oxygen by vacuum melting, mass spectrometry, and fast-neutron activation are done without etching the surface of the sample before measuring the activity. It also shows the data obtained by NAA and ^{3}He activation with a careful removal of contaminated surface layers of a thickness of at least 10 microns (of the same samples). The data indicate a good agreement between the two methods of activation analysis with the removal of the contaminated layer. At the same time, the discrepancy between the data received by "traditional" methods (vacuum-extraction, coulonometry) with the similar removal of the surface layer and two activation methods with the removal of the oxidized layer is almost two orders of magnitude for the same samples. This is strong evidence in favor of the use of highly sensitive methods of analysis for determining the oxygen concentration below 10^{-3}%. The difference in the slope of the kinetic curves for different modes of the melting chamber warm-up due to the influence of the residual atmosphere (especially moisture). The results of this work are shown that the mechanism of the oxygen removal from the liquid niobium remains constant over a wide range of oxygen concentrations from 10^{-2}% to 10^{-5}%. Data obtained earlier by other authors on changes in the mechanism are due, apparently, to the incorrect determination of the oxygen content by conventional analytical methods.

When the flow of the two competing processes (desorption of oxide layers of niobium and carbon monoxide from the surface of the molten zone) have a situation where at a certain critical oxygen content the carbon removal slows down sharply, but oxygen in the form of oxides of niobium continues to be desorbed. As a result of the multiple EBFZM of niobium

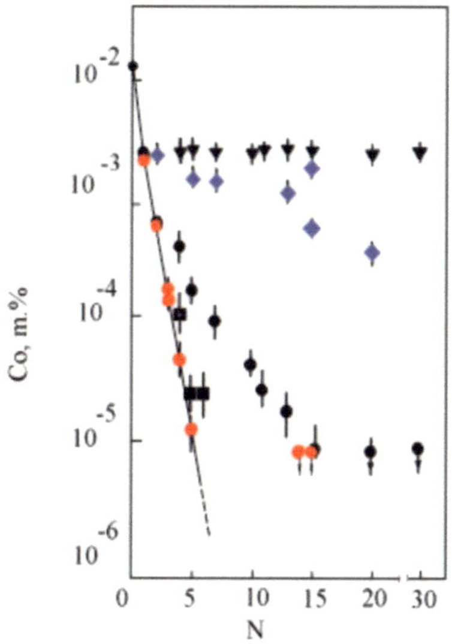

Figure 9. Dependence of the oxygen concentration on the stationary part of niobium single crystals of the number of zone runs; single samples of every run were analyzed by several analyzing techniques: ○,●– NAA (red & black, two different regimes of heating-up the chamber), ■ – ^{3}He, ▼- vacuum-extraction, samples after standard cleaning; ◊ - vacuum-extraction (blue), samples after careful cleaning (removal of a layer of at least 10 microns).

doped with carbon and oxygen, it is found that at reducing the oxygen content in the range of $6 \times 10^{-3}\%$ and $1.5 \times 10^{-4}\%$ the nature of the interaction of carbon and oxygen changes: the carbon content tends to a limiting value for the initial contents of both interstitials and the faster, the closer to the "critical" range is the oxygen content after this zone run. To reveal limits of the critical range of oxygen concentrations with greater accuracy is not yet possible because of the discrete nature of the vacuum zone melting, and the impossibility of fixing a continuous fall of the contents of both interstitials. The dependence of the degree of decarburization during vacuum zone melting from exceeding the initial oxygen concentration above the "critical" range for the crystals with the same initial value for both interstitials has been studied, and it is shown that at the transition to higher absolute contents of both interstitials, the degree of decarburization increased by more than two orders of magnitude. It is found that the decarburization efficiency of niobium depends primarily not on the ratio of the initial ratio of O/C, but on the difference between the initial and critical oxygen concentrations: the higher the difference, the more intense decarburization proceeds. The data obtained experimentally confirm the validity of the proposed model of the interaction of carbon and oxygen: when achieving the certain critical oxygen concentration, the carbon removal rate slows dramatically and establish the limiting concentration of carbon in niobium.

In similar experiments on molybdenum doped with oxygen and carbon in the wide concentration range, it is shown that during HF vacuum levitation the contents of both impurities are reduced to $5 \times 10^{-5}\%$ and $10^{-6}\%$, respectively, in approximately 15 seconds. This fact, on the one hand, also demonstrates the validity of the model proposed concerning the removal of the interstitials by forming CO and MoO_2. On the other hand, these experiments indicate that there is no fundamental obstacle to obtaining molybdenum, substantially free of carbon and oxygen in the melt in vacuum.

3. Technological plasticity of unalloyed high-purity refractory metals produced by EBM

Practical applications of refractory metals, produced by means of EBM, are limited due to the intergranular brittleness of cast polycrystalline ingots (especially of molybdenum and tungsten), although these metals in the single-crystalline state are sufficiently plastic in a wide temperature range. It is known that interstitials are almost always present in cast refractory metals as a second phase and contribute to the appearance of cracks at grain boundaries. The low technological plasticity and structural strength of refractory metals are particularly bad at medium and low temperatures. Studies [56–58] are shown that molybdenum of vacuum melting can be of the appreciable technological plasticity if the interstitials contents are lower than $1 \times 10^{-4}\%$ of each element. These data for many years were a kind of a truth for researchers who worked on this problem. The real prospect to get molybdenum with the low content of interstitials arose with an advent of industrial vacuum electron-beam melting (EBM). However, ingots of unalloyed molybdenum of EBM, despite the relatively low content of interstitials, have still the low technological plasticity because of the increased segregation of interstitials in clusters at grain boundaries, which exceeds the average concentration of interstitials in the volume of ingots for several orders of magnitude. Since it is impossible to achieve the greater technological plasticity by vacuum melting only, the researchers direct their efforts towards the creation of alloys in which the fracturing effect of interstitials has been decreased by doping. Additives added during melting prevent the segregation of interstitials along the boundaries and bind residual interstitials in the dispersed second phase. Well-known dopants are metals of Groups 4, 5, 8, such as boron, aluminum, carbon, and rare earth metals. Currently, there is a wide range of cast molybdenum and tungsten alloys of the satisfactory technological plasticity. However, in recent decades there is a sharp increase in the interest of undoped high-purity refractory metals. This is due to the fast development of industrial areas of the critical application. Doped alloys are almost completely excluded from the practice of the thin-film metallization in microelectronics. Thus, the purity and technological plasticity of the EB-melted undoped polycrystalline molybdenum and tungsten attract attention.

3.1. Molybdenum

The low plasticity (or the high intergranular brittleness) of coarse-grained, electron-beam melted molybdenum is a well-known feature. It is believed that the brittleness may be caused by the segregation of interstitials (oxygen, carbon) at grain boundaries. A common way to avoid the embrittlement of this material consists of the addition of deoxidants or modifiers such as carbon, boron, and other elements to the feedstock before or during electron-beam melting (EBM) to produce the fine-grained macrostructure [59–62]. We have studied the possibility of improving the plasticity of molybdenum by subjecting it to high-temperature processing. Three ways are tested for preparing pure molybdenum ingots. The first way consists of the single EBM run at the melting rate, at which ingots have the smallest grain size or the fine-grain structure. The second way is the multiple EBM run, and the third way is the duplex process consisting of electron-beam melting and vacuum arc melting (VAM). Molybdenum ingots, 80 mm in diameter and 1300 mm in length, are melted in an electron-beam furnace in a vacuum of 1×10^{-3} Pa with melting rates ranging from 0.2 to 2 kg min^{-1} [60-63]. The number of runs ranges from one to five with the melting rate ensuring the maintenance of the melt in the mould (crystallizer) at the appropriate temperature. The mean power of the EBM is 100 kW at an accelerating voltage of up to 25 kV and an emission current of 4 A. The melting of molybdenum ingots of 140 mm in diameter and 350 mm in length is performed in the VAM set-up in a vacuum of ~10^{-2} Pa at a mean voltage of 35÷40 V and an arc current of 4.5÷5 kA. Cylindrical specimens are cut out of each ingot for chemical analysis, metallography, and mechanical testing. For microfractography, flat specimens (10 mm × 10 mm × 50 mm) are cut out normal to the ingot axis. The fracture plane of the specimens is parallel to the ingot axis. The mechanical testing of the specimens is carried out by a conventional technique in an Instron machine. Light and electron scanning microscopes are used for microfractography and metallography. The grain size is determined by directed secants on the transverse and longitudinal macrosections of polycrystalline ingots. The texture of ingots is studied by observing X-ray diffraction patterns. The behavior of the material under processing conditions (pressing and forging of ingots, rolling of billets) is tested in a standard way at 900–1300°C, with the ingots heated in induction furnaces in vacuum or in a hydrogen atmosphere. Attempts to vary the melting rate over a wide range during the run in the EBM set-up of the PM feedstock are unsuccessful because the melting rate is less than 0.5÷0.6 kg min^{-1} and is limited by intensive outgassing of the feedstock and liquid metal. The EBM ingots of the first run have a coarse structure and many metallurgical macrodefects: gas voids, unmelted PM fragments, and severe fissures (Figure 10).

Figure 10. Gas void at the fracture surface of the molybdenum ingot (In the vicinity of a PM fragment, the center).

As usual, to produce the ingot of good metallurgical quality the second EBM run is performed at a higher melting rate. The melting rate has a strong effect on the macrostructure: the mean grain size is 5–6 mm for the low melting rate and 1–2 mm for the high rate and the grain lengths are respectively 100 mm and 10 mm or less (Figure 11).

The texture analysis shows that the grains in all ingots have a longitudinal axis close to the (001) orientation. With successive runs carried out at the rate of ~1 kg min^{-1} the average size of

Figure 11. Macrostructure of the molybdenum ingot of 1m length after the EBM run at 0.2 kg min^{-1} (left part) and 2.0 kg min^{-1} (right part).

grains increases slightly: it is 2.2 mm after the second run and 2.7 mm after the fifth. Figure 12 depicts the grain size distribution for two macrostructures of the molybdenum ingot after two runs carried out at the rates of 0.2 kg min^{-1} and 2 kg min^{-1}.

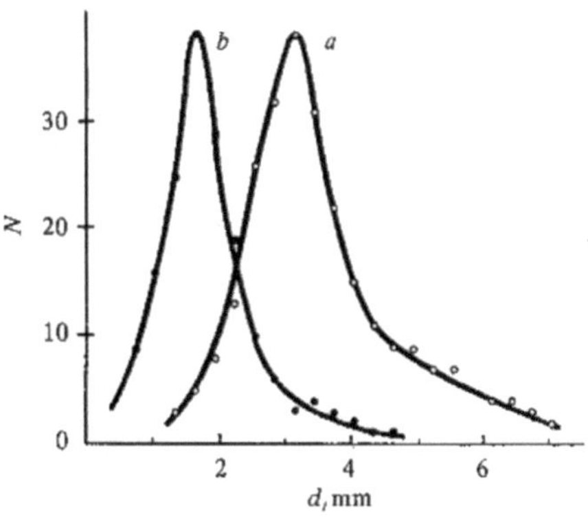

Figure 12. Grain size distribution in the molybdenum ingot after the second run by EBM at the rate (a) 0.2 kg min^{-1} and (b) 2.0 kg min^{-1}.

The etch pit density in pure molybdenum ingots decreased as the number of melts increased from one to five. The grain boundaries were less strongly etched and became thinner with increasing number of runs. As seen from Table 8, after the first and second runs all metallic impurities were near the detection limit (0.01÷0.1 ppm) of the ICP mass-spectroscopy. The tungsten content was nearly constant (~10 ppm) in the course of the runs. After the fifth run all metallic impurities and interstitials in molybdenum were below the detection limit of analytical techniques used.

In our version of the duplex EBM-VAM technique, the EBM ingots are used as an initial material and have been melted in the vacuum arc furnace with a melting rate of 2.1–2.3 kg min^{-1}. The main feature of the VAM technique is that the liquid metal is stirred by the electro-magnetic field of a solenoid mounted outside of the water-cooled crystallizer. Figure 13 depicts the grain size distribution in two molybdenum ingots produced with different stirring currents. It is seen that the macrostructure strongly depends on the stirring current. At low

http://dx.doi.org/10.5772/59397

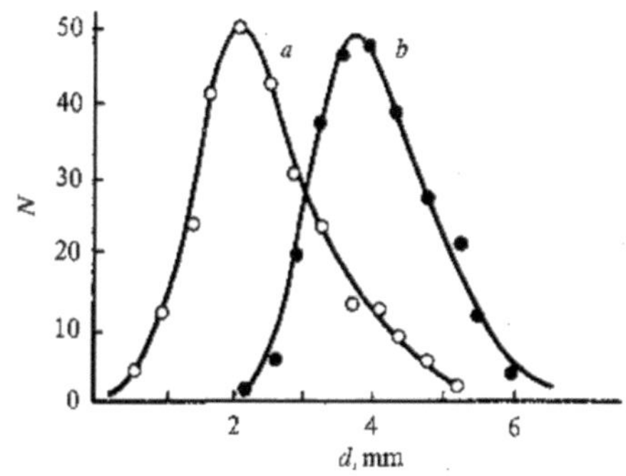

Figure 13. Grain size distribution in the molybdenum ingots prepared by duplex melting with different stirring current: (a) 1 A, (b) 4.5 A.

Number of runs	Impurity content, 10^{-4}% m.							
	C	O	N	Ni	Fe	Ti	Ta	W
PM feedstock	30	40	10	2	100	5	1	20
Single run	8	2	1	<0.3	1	<0.3	1	10
Double run	8	0.5	<1	<0.3	<1	<0.3	<1	30
Fivefold run	8	<0.5	<1	<0.3	<1	<0.3	<1	30

Table 8. Impurity content in EB-melted molybdenum depending on the number of runs

stirring current (1–1.5 A) the macrostructure is characterized by grains up to 60 mm long with a mean size of 4 mm, inclined to the ingot axis by 15°. At higher stirring currents (up to 5 A) the grains are smaller (~2 mm) and shorter (~15–20 mm). The carbon and oxygen contents in molybdenum produced by the duplex melting are shown in Table 9: the higher the stirring intensity, the purer the metal. In any case, the cast molybdenum is not contaminated during arc-melting at a lower vacuum.

The normal cross-section of the molybdenum ingot of 80 mm in diameter, produced by the duplex EBM+VAM, is shown in Figure 14. Here, it can be seen the homogeneous structure with small grains, except the subsurface layer. Microfractography showed that the specimens from all ingots (EBM, EBM + VAM) have a mixed transcrystalline/intercrystalline fracture with micro-voids and inclusions on the fracture surface (Figure 15a). The relief of the fracture surface varies in the vicinity of the inclusions, indicating their influence on the plastic flow. Cracks originate near the voids or flat precipitates; Figure 15b depicts a void presumably formed as a result of the breakaway of a carbide particle.

The high-temperature plastic deformation of molybdenum ingots subjected to multiple EBM or duplex melting (except for ingots after the first EBM of the PM feedstock) occur without failure of the specimens. About 20% of the ingots tested after the first EBM run fail because of cracks during manufacturing.

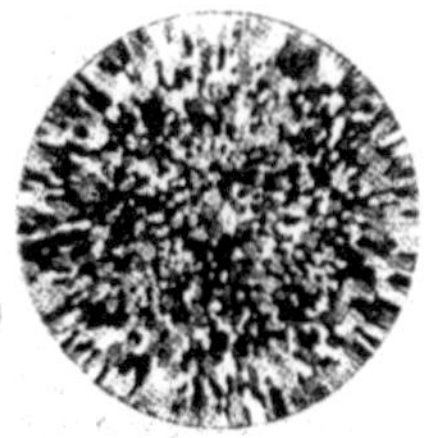

Figure 14. Macrostructure of the molybdenum ingot produced by duplex melting with the highest stirring current.

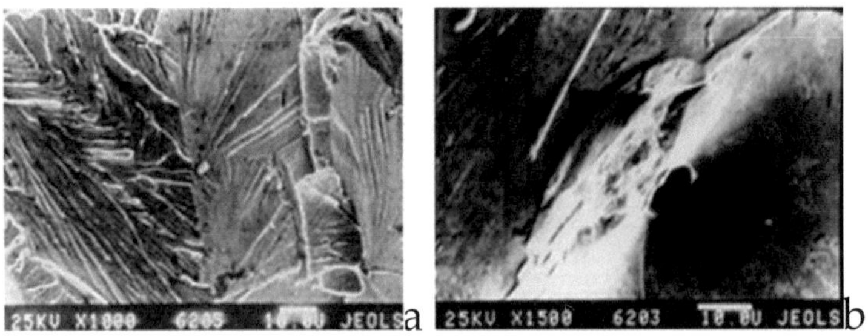

Figure 15. Fracture surfaces of molybdenum: (a) mixed fracture (double EBM run); (b) void, formed after breakaway of carbide (EBM+VAM).

The plastic deformation of 75–82% and the temperature of 1200–1300°C are chosen to produce billets with uniform grains. Sheets of 0.3 mm thickness are prepared by hot rolling. Mechanical tests show that molybdenum ingots are ductile at all stages of the high-temperature fabrication. The mechanical properties of annealed and unannealed rolled molybdenum specimens are presented in Table 9. The temperature dependence of the mechanical properties of 0.3 mm thick annealed sheets is shown in Figure 16 over the temperature interval 200°C to –192°C. The mechanical testing reveals good ductility and high strength of coarse polycrystalline ingots of molybdenum melted in the electron-beam or vacuum arc set-ups. Although information on the behavior of carbon and oxygen during vacuum melting of molybdenum is rather scant, it is known that oxygen can be removed as oxygen atoms, carbon oxides, and molybdenum oxides, but carbon can be removed only as carbon oxides. At low carbon and oxygen contents the interstitials in molybdenum behave independently: the carbon content remains constant in the course of repeated EBM, whereas oxygen content sharply decreases. Thus the vacuum deoxidation of molybdenum (or self-deoxidation), in which the oxygen removal occurs owing to the evaporation of carbon oxides and volatile molybdenum oxides can be used for the production of molybdenum ingots of high purity and good mechanical properties.

The study of ingots produced by the double EBM run reveal the significant effect of the melting rate on the macrostructure of the undoped high-purity molybdenum. The average grain size reduces from 6 to 2 mm (about 3 times) with an increase in the melting rate of 0.2–2 kg/min (i.e., 10 times). The average grain length decreases respectively from 100 to 10 mm. The ingots produced at the melting rate of 0.2 kg/min contain columnar grains, which are inclined from the axis of the ingot. The texture analysis reveals that all ingots contained grains with the longitudinal axis close to the [001] direction.

Stirring current, A	Oxygen content, 1×10^{-4} m%		Carbon content, 1×10^{-4} m%	
	EBM	EBM-VAM	EBM	EBM-VAM
1	2	9	8	8
3	5	6	8	8
5	4	4	8	8

Table 9. Oxygen and carbon contents after EBM and EBM+VAM

Annealing temperature, °C, 30 min	σ_b, MPa	δ, %
Unannealed	990	7.5
800	835	13
900	740	22
1000	630	27
1100	640	29

Table 10. Ultimate tensile strength, σ_b, and elongation, δ, of molybdenum purified by the duplex process EBM+VAM

Increasing the number of EBM runs from 2 to 5 at the rate of melting of 1–1.2 kg/min results in an increase of the average grain size up to 2.7 mm (after five runs) at a constant length of 10–20 mm. Further reduction of impurities is able to establish only by indirect metallographic methods, since the content of almost all impurities after two runs is beyond the detection of the analytical methods. Fractographic tests show that after two runs all the samples have the mixed fracture with a predominance of the transgranular fracture. Samples of one run have the intergranular fracture behavior and on fracture surfaces are observed precipitates (carbides). Cast ingots of undoped high-purity molybdenum of all five runs are plastically deformed by high-temperature pressing using the standard technological scheme adopted for ingots of doped molybdenum. The processing cycle includes heating to 1300°C in vacuum, hot pressing at the high-speed press with a force of 670 tons, recrystallization annealing in a hydrogen atmosphere at 1300°C, hot (900–1300°C) and warm (up to 300°C) rolling. Processing ingots by two to five runs is held normally, without cracking. On ingots of unalloyed high-purity molybdenum of the single run at the stage of hot pressing the cracks appear. The deformation degree is 75–82 %, and the recrystallization annealing temperature is adjusted so that it is possible to obtain a semi-integrated rolling feed with the uniform grains because unalloyed molybdenum has the low recrystallization temperature of about 1000°C (Table 10). The tensile strength of annealed ribbons of undoped high-purity molybdenum is 72 kg mm^{-2}, the elongation of 15.7%. Processing cast ingots of undoped high-purity molybdenum produced by EBM show its high adaptability, which is almost entirely dependent on its purity and lack of large and elongated columnar grains with a poor cohesion. The high technological plasticity of undoped high-purity molybdenum allow producing workpieces of rectangular form or flat bars for deformation processing, cut from cylindrical ingots or melted with a special "flat" mold. Hot-rolling flat bars to a sheet with a thickness of 6.5 mm is similar to the procedure adopted for rolling pressed billets. The use of the flat ingots of undoped high-purity molybdenum allows to eliminate one of the deformation runs and avoid the uncontrolled contamination of the metal and metal losses arising in the preparation of feeds for pressing and rolling. During the duplex process ingots of undoped high-purity molybdenum of the single EBM run are melted in the VAM set-up (vacuum, argon) with a consumable electrode and electromagnetic stirring of the melt. In this case, the consumable electrode is an ingot of high-purity unalloyed molybdenum of the smaller diameter (60 mm) produced by EBM in the EMO-250 set-up (250 kW, axial electron gun, vacuum). The run is carried out with a rate of 2–3 kg/min, with stirring the melt by electromagnetic field of the solenoid. At currents of 1–1.4 A at the solenoid the macrostructure is characterized by grains of up to 60 mm with an average grain size across the eye ~4 mm. At stirring currents up to 5 A the grains lengths are decreased to 15–20 mm, and the average grain size in diameter is 2 mm. The parameters of the electric arc remains constant during the whole run. According to the microstructural analysis the recrystallization of undoped highly pure molybdenum occurs in the temperature range 900–1100^{0}C. The results of the mechanical testing of sheet samples of undoped high-purity molybdenum of EBM+VAM have a quite satisfactory performance even at low temperatures from -192^{0}C up to 200^{0}C (Figure 16).

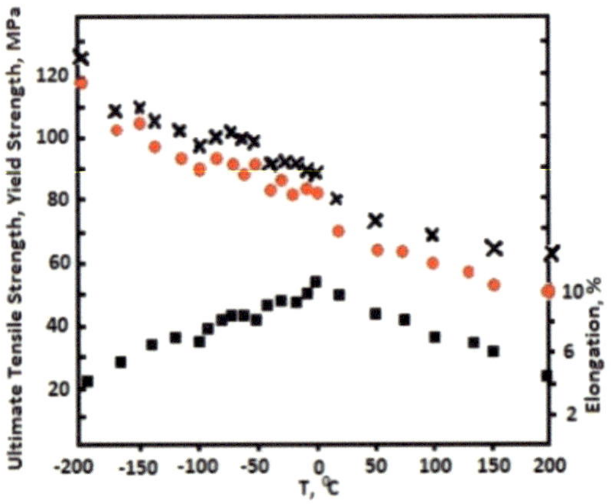

Figure 16. Temperature dependence of the ultimate tensile strength, σ_b (black crosses), yield strength, $\sigma_{0.2}$(red circles), and elongation, δ (black squares).

3.2. Tungsten

An increase in purity of tungsten can open the possibility of increasing the technological plasticity. Purifying polycrystalline tungsten from gas-forming interstitials, especially oxygen and carbon, is considered as one of the most effective ways to improve its ductility. The vacuum melting techniques such as electron-beam vacuum melting and electric arc vacuum melting give additional excellent possibilities (especially, refractory metals of high purity) to produce many tungsten products of the higher quality compared with powder metallurgy (PM) products [63-65]. However, in the case of tungsten, the high purity alone cannot give polycrystalline tungsten ingots of sufficient technological ductility. The defects, such as voids, second-phase inclusions, PM-fragments, and inhomogeneous coarse-grained structures, are only the part of the problem that should be solved before any technology could be recommended for vacuum melting, pressing, extrusion, and rolling the thin sheets and other tungsten products. Apart of this, during plastic deformation and other thermal treatments the contamination of tungsten with hydrogen and oxygen is quite possible, which can introduce different negative effects. This research is devoted to the study of an influence of some structural defects on the plasticity of the high-purity tungsten ingots [63–65]. The tungsten ingots are produced from the relatively pure PM feeds by double vacuum melting consisting of the electron beam melting (the first run) and electric arc-vacuum melting (the second run) with the intensive electromagnetic mixing of the melt. As can be seen from Table 11, tungsten is purified mainly at the first EB-run. The second arc-vacuum melting is important for a better structure homogeneity and the decreasing the tungsten ingot porosity. It seems that the structural quality and the technological plasticity of the tungsten ingots could be improved by studying the melting and plastic deformation processes. The technology is developed for obtaining cast tungsten ingots and their subsequent pressing and rolling.

	Mo	O	C	Si	Ti	P	S	Al	N	Co	Fe
PM rods	50	30	20	10	10	20	10	10	5	2	2
Single run	50	0.5	10	0.2	0.3	0.05	0.05	0.1	0.9	0.03	0.1
Double run	50	0.5	10	0.2	0.3	0.05	0.05	0.1	0.3	0.03	0.1
	F	**Hf**	**Cr**	**Ca**	**Ta**	**Cl**	**Mg**	**B**	**Ni**	**Na**	**K**
PM rods	-	1	1	1	1	1	1	1	-	1	1
Single run	0.3	0.3	0.1	0.1	1	0.05	0.03	0.01	0.1	0.03	0.03
Double run	0.3	0.3	0.1	0.1	1	0.05	0.03	0.01	0.1	0.03	0.03

Table 11. Chemical composition of tungsten ingots (1×10^{-4}m%).

The ingots for pressing and extrusion are heated in a HF induction vacuum furnace. Hot pressing and extrusion of the cast polycrystalline tungsten ingots are produced with the relative deformation of 70–78%. Then billets for rolling are heated in a hydrogen furnace with a dew point of hydrogen lower than –30°C. Hot rolling of billets is done in a two-high rolling mill equipped with cast iron rolls of 300 mm in diameter and with a rolling rate of 20 m min^{-1}. The density of tungsten is determined by hydrostatic method. The contents of impurities, especially interstitials, are analyzed by mass spectral, vacuum extraction, and nuclear physical methods. The pores and carbide particles of 1 micron or greater are determined by metallographic methods. The mechanical grinding and polishing with diamond powders are used to prepare specimens to study the structure. After electropolishing in an electrolyte containing soda, chemical etching is carried out by a specific acidic etchant. To avoid overetching of grain boundaries and pores and crumbling of carbides this technique is optimal. Electron scanning microscopy is used to study the pores at large magnifications. The microanalysis of carbon in the carbide particles is conducted in X-ray microanalyser.

In the ingots of 120 mm in diameter the pores are found in the layers up to a depth of 15 mm from the surface (Figure 17). Exclusions accounted only for ingots of an average density (determined by a specific hydrostatic method) of less than 18.9 g/cm^3 (99% of theoretical density of tungsten), the lower density ingots are unsuitable for further processing and referred as the "low" density ingots. In this case, the pores of 5–100 microns are detected through almost all of the bulk of ingots. In the subsurface layers the pores are up to several millimeters. Ingots with a density higher than 18.9 g cm^{-3} are referred to as the "normal" density ingots and contained pores only in the subsurface layers, but at a greater depth the pores are absent. The density estimates obtained by metallography are in a good agreement with results of a hydrostatic density technique. In high-purity tungsten ingots of double EB melting (double run), the sizes of the peripheral grains are 2–2.5 times bigger than in ingots of tungsten doped with carbon (70 ppm). In the center of ingots of the double run the grains are the largest. Thus, it is revealed that grain sizes in the high-purity tungsten ingots are increased from the peripheral area to the center; however, grain sizes in the tungsten ingot doped with carbon, distribute almost uniformly.

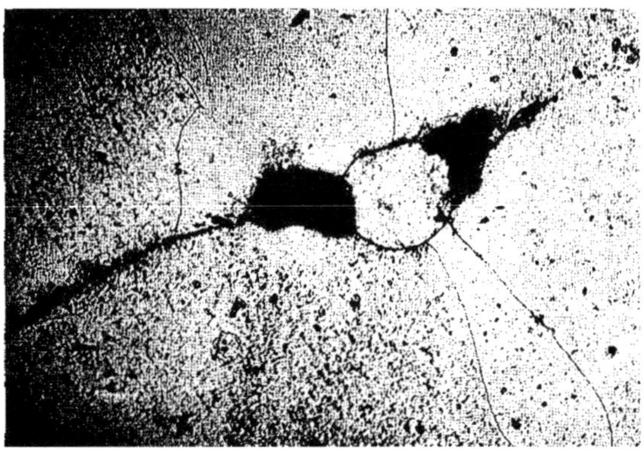

Figure 17. Pores in a tungsten ingot of the "low" density (× 125).

It is found that an absence of pores, cracks, and other structural defects is very important for high-temperature plastic deformation of tungsten. The largest number of defects in the form of pores occurs in the undoped high-purity tungsten ingots of the single EB run, ingots of the double EB run – substantially smaller, ingots of the carbon-doped tungsten produced by the EB-run – the smallest. During the hot plastic deformation of tungsten ingots, the pores change their shape according to the ingot formation. Ingots of undoped high-purity tungsten with highest soundness and technological plasticity are produced by the duplex process consisting of the EB melted ingots (several electrodes) followed by the run of these ingot-electrodes in the VAM setup with the electromagnetic stirring of the melt. It is much more difficult to observe pores in samples obtained after hot pressing the ingots since as a result of deformation their cross-section size is drastically reduced (Figure 18) and the density of ingots is increased. Among the *bcc* metals tungsten stands out because of its high brittle–ductile transition temperature (usually T_{bd} = 400–600°C). At a temperature below T_{bd} the rolling leads to the brittle fracture of tungsten: for example, the rolling tungsten sheets at 400–600°C should therefore be considered as dangerous. At different stages of the rolling, according the technological regulations, some temperature losses are quite possible. For example, the tungsten sheets should be taken from the hydrogen furnace to the rollers, which usually takes 5–10 s. This procedure leads to typical temperature losses (due to the convective and radiative heat exchange with the ambient medium), which do not exceed 50°C [66]. The tensile strength of pure tungsten above 1100°C is about 300 MPa [67]. Our estimations show that for such a large one-step 60% deformation, the deformation part of the temperature increment is about 70°C. The thermal contact of the hot sheet with the cold rolls should significantly decrease the temperature increase due to deformation. It seems that the most serious temperature decrease occur due to the thermal conduction at the contact between the tungsten sheets and the rolls during rolling.

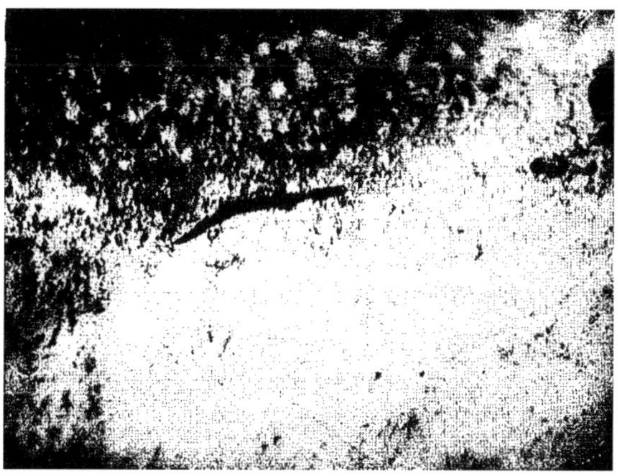

Figure 18. Pores in pressed tungsten prepared from the "low"density tungsten ingot (× 630). The plane is perpendicular to the pressing axis.

A relationship between the sheet thickness and the averaged ΔT is obtained in [68], which allows to estimate the cooling of tungsten billet on the rolls during rolling as a function of the initial temperature (T_w) for a roll temperature, T_r :

$$\Delta T = \frac{4}{H+h}\left(\frac{L}{V_r}\right)^{0.5} \frac{T_w - T_r}{1/a_w^{\ 0.5} + 1/a_r^{\ 0.5}\left(c_w \rho_w / c_r \rho_r\right)}$$

where a_r and a_w are the heat conductivities of the cast iron rolls ($1.44x10^{-5} m^2 s^{-1}$) and tungsten at 1400°C ($3.8 \times 10^{-5} m^2 s^{-1}$) respectively. The quantities ρ_w and ρ_r are the densities of tungsten and rolls, c_w and c_r the respective heat capacities, L the length of the deformation site along the rolling direction, V_r the rolling rate, H and h the thicknesses of the sheet before and after rolling respectively. The increase in the length L of the tungsten–rolls contact as compared with $\{R(H-h)\}^{1/2}$ is not taken into account in estimating ΔT. This increase is aroused from elastic flattening of the rolls, especially noticeable during rolling of tungsten. The differences in the heat transfer to the rolls due to formation of the oxide film on tungsten, having an opposite influence on the ingot cooling, are omitted from consideration. Table 12 shows that the temperature losses approach the critical value ΔT=700°C as rolling sheets of 2 mm thick (H+h = 4÷5 mm).

Rolling reduction per passage, (H–h) mm	**(H+h),mm**					
	3	4	5	6	8	10
0.2	643	482	386	321	241	193
0.4	766	574	460	383	287	230
0.6	857	643	514	428	321	257
0.8	904	678	543	452	339	271
1.0	952	714	571	476	357	286

Table 12. Estimation of the mean temperature decrease (°C) on rolling of tungsten heated to 1400°C for different thicknesses of the sheet at the inlet (H) and outlet (h) of the rolling gap

From the technological practice is known that it is at this thickness the tungsten sheets are fractured more often. If an initial microstructure of the tungsten ingots does not guarantee decreasing T_{bd}, the rolling the tungsten sheets below an actual rolling temperature makes them brittle and leads to fracture. It is necessary that the decreasing ductile–brittle transition temperature follows the decreasing rolling temperature as the thickness decreases. So, if T_{bd} would be lower the rolling temperature, the rolling would be still successful: the highly pure tungsten of the "normal" density could be rolled to 1.2–1.5 mm. The sheets from the "low" density tungsten ingots at the same conditions would fracture at 2–4 mm that look like the critical thicknesses. Table 12 demonstrates the mean temperatures at which final passages take place differ by about 200°C. For the "normal" density tungsten (H+h) is less than 3 mm whereas for the "low" density tungsten (H+h) is more than 4 mm (with reference to initial tungsten ingots).

The temperature T_{bd} of high purity tungsten of the "normal" density is for 200°C lower than that for tungsten of the "low" density ingots. Cracks and voids in fractured tungsten sheets of the "low" density look very different (Figure 19). The specimens for scanning electron microscopy are prepared in the plane perpendicular to the rolling plane. The preparation is done without electropolishing or etching. Thus traces of scratches are even seen in Figure 19 in the direction perpendicular to the rolling. The voids and cracks in the cross-section are typically of 1–5 microns. It seems as if the cracks are initiated from one point. It should be added that even the rolling is done alternately with anneals, the porosity inherited from the "low" density tungsten ingot could not eliminated. In this respect vacuum-melted tungsten is inferior to that obtained by powder metallurgy. In the latter the small grains and pores of 10 microns in the sheets can be completely sintered of at intermediate anneals. It is absolutely impossible in the coarse-grained vacuum-melted tungsten, which has typical grain sizes of several millimeters. The pores in the "low" density tungsten act like crack nuclei or stress concentrators. To save this possibility, they should be capable before reaching the critical thickness of about 2 mm. Another factor of strengthening and the growth of defects in the "low" density tungsten is hydrogen out of atmosphere of the hydrogen furnace. Hydrogen penetrates very easily into tungsten: The diffusion coefficients (D) of hydrogen are equal to 2 × $10^{-5}$to 1.9 × 10^{-4} cm^2 s^{-1} at 900K and 1.6 × 10^{-4} to 2.6 × 10^{-3} cm^2 s^{-1} at 1600 K [69]. After 2 min of heating in the hydrogen furnace, hydrogen penetrates on 0.5 mm at 900 K and 1.4 mm at 1600 K. Hereafter the depth of diffusion is estimated as $(D_\tau)^{1/2}$, corresponding approximately to 50% of the concentration C_0, whereas $2(D_\tau)^{1/2}$ is usually used, corresponding to the lower concentration $0.3C_0$. The equilibrium concentration of hydrogen is 3.3 × 10^{-6} at.% at 1200K and 4.0 × 10^{-5} at.% at 1600 K if the hydrogen pressure would be close to 1 atm. As the sheets have been cooled during transportation to the rolling mill or upon rolling, the tungsten becomes oversaturated with hydrogen. Hydrogen travels to the free surfaces and pores, which act as sinks. It is especially pronounced if thickness is less than 5 mm (Table 12, H+h < 10 mm) and the cooling increases dramatically.

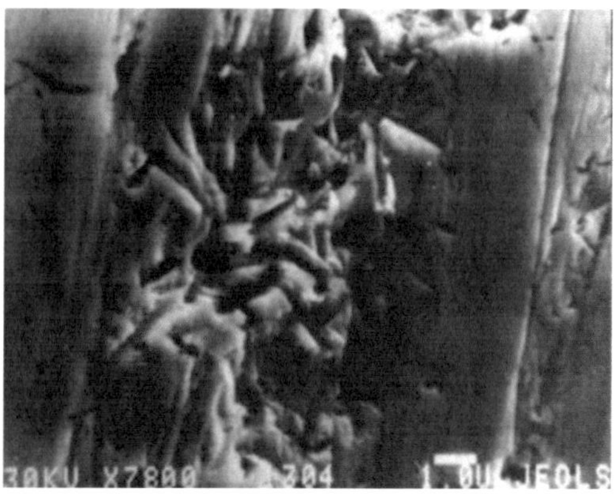

Figure 19. Cracks and pores in hot-rolled sheets of the "low" density tungsten (× 7800). The plane is perpendicular to the rolling direction.

The estimate shows that on cooling for the time period $\tau = 2$ s and starting from the temperature $T_1 = 1400°C$ going to $T_2 = 1200°C$, the equilibrium solubility of hydrogen in tungsten decreases by a value $\Delta c = 4 \times 10^{-5}$ at.%. The characteristic hydrogen diffusion distance during this time is $(D_\tau)^{1/2}$ ~ 140 microns (the diffusion coefficient at $T_2 = 1200°C$). At $\alpha \leq 1$ (part of hydrogen), a site of radius $(D_\tau)^{1/2}$ would collect molecular hydrogen in a pore of a volume V_p, thus its pressure would be:

$$p = \frac{\alpha \Delta c 4\pi \left(D\tau\right)^{3/2}}{3\mu_w V_p 2} \rho_w \left(T_2 + 273\right) R.$$

where μ_w is the tungsten atomic weight, 2 the atoms in the molecule of hydrogen, ρ_w the density of tungsten, and R the gas constant. A pore with a radius of 4 microns and $\alpha = 0.1$, so p~12 at. In the case of $\alpha = 1$ and the radius $2(D_r)^{1/2}$ of the site, thus p ~ 960 at. It is necessary to know that when the billets or sheets getting smaller during plastic deformation, diffusion, and collection of hydrogen in pores can be fastened. As a result of this process the hydrogen pressure would become much higher than the estimated one. At unloading when the workpiece comes out of rolls, the pressure in pores can be so high that it can be like a source of the explosive initiation and abrupt growth of cracks. The high potential ability of hydrogen to produce significant deformation due to the large and closed pores in tungsten sheets is indirectly confirmed by swelling observed after annealing at 1500°C over and over again (Figure 20). It is the well-known experimental and technological fact: large pores of the initial tungsten ingot are elongated upon rolling and begin to swell. Owing to an increase in pressure and softening of tungsten, the swelling may occur upon heating or resulting from formation of the oversaturated solid solution and collection of hydrogen in pores at temperatures at which tungsten is sufficiently plastic upon cooling. An important feature of the hot rolling of tungsten is the real possibility of the oxygen penetration from surface oxides into the sheets, despite the fact that the sheets are heated before rolling in a reductive hydrogen atmosphere. From our experiments, the average oxygen contents, which can be achieved during technological procedures on tungsten, are usually within (0.2–5) × 10^{-4}% with respect to the thickness of the sheets; however, the oxygen contents also depend on the heating temperature and an exposition at which the heated sheets were in air. In this situation the data available concern the diffusion of oxygen in tungsten. It should be noted that tungsten samples (more than 30 samples are examined) with the mean content with respect to thickness higher than 1 × 10^{-4} % contained cracks formed either upon rolling or during mechanical preparing the specimen for the neutron activation analysis of the oxygen content. We have also studied the carbides distribution in the cast tungsten ingots which is rather non-uniform: at the volumetric carbide content of 0.017–0.055% (the measurements are made using a point technique, several specimens are studied, 25–50 fields of vision, 256 points in each one). There are two different kinds of the porosity defects: areas containing several carbide particles and sites enriched in oxygen. It is found that about 30% of the carbides observed are located at grain boundaries, the remainder being present inside the grains. The carbide particle sizes typically range from 1 to 10 microns. The particles observed are actually carbides, which is supported by two reasons. At first, the microhardness is 13.620 MPa, whereas ones of a tungsten matrix are 4.020–4.120 MPa.

The particles contained larger amounts of carbon as shown by X-ray microanalyser. It is revealed from estimates of a volume fraction of the observed particles that contain carbon satisfactorily agreed with the results of the chemical analysis of carbon. It is shown that carbon contents in the solid solution (carbides) are about 1 × 10^{-4}% (approximately 10% of the total carbon content).

Figure 20. Swelling of a hot-rolled tungsten sheet from the 'low" density ingot, annealing for 2 h in hydrogen at 1500°C: (a) ingot surface (×2); (b) transverse section, subsurface discontinuities is seen (×80); (c) pore with hydrogen, looks like elongated "bubble" after annealing, splitting cracks are seen (×100).

4. Key findings

1. A model is proposed within the framework of which specific features of the interaction between interstitials (carbon, oxygen) during high-temperature annealing of refractory metals in vacuum are considered. The limiting carbon concentration that can be attained during a sufficiently prolonged annealing process has been found to depend on the initial carbon/oxygen ratio in the specimen, the specimen geometry, size, and temperature. It has been established that there exists a certain critical concentration of oxygen beginning with which the oxygen effect becomes noticeable and is the more pronounced as the characteristic size of the specimen decreases. A procedure for determining the sticking probability and the effective desorption rate constant for carbon oxide molecules is proposed. The existence of a characteristic critical carbon concentration is shown, above which it can be "burn off" from the metal. The optimal conditions for decarburization annealing of refractory metals (molybdenum, tungsten) are found according to the effective partial pressure of carbon monoxide in the gaseous phase. In a wide range of contents of both interstitials, the kinetics of the interaction processes in liquid niobium, molybdenum, and tungsten in vacuum is studied using the high-frequency levitation and electron-beam zone melting. The direct experimental proofs of the validity of the proposed model and basic predictions are received. The possibility of obtaining refractory metals with the lowest contents of carbon and oxygen is shown: during experiments contents of carbon and oxygen were at the detection limits of activation analysis methods, with the detection limits of the fast neutron activation for oxygen and deuteron activation for carbon being $10^{-5}\%$ and $10^{-6}\%$, respectively. The regularities and mechanisms of the removal of carbon from the molten niobium, molybdenum, and tungsten at high temperatures in vacuum are studied. The basis is developed on the connection between their satisfactory technological plasticity and the interstitials segregation on grain boundaries. It is also shown that the connection is linear, but rather complicated.

2. The relationship between the macrostructure and plasticity of molybdenum has been investigated. It is shown that undoped polycrystalline molybdenum of the high purity, produced by multiple electron-beam melting (EBM) or the duplex process consisting

of EBM and vacuum arc melting (VAM) has the satisfactory technological plasticity when subjected to high-temperature processes (pressing, rolling, forging). This feature of this material appears to be connected with the low content of interstitials on the intergranular surfaces and an absence of macrodefects in ingots.

3. The possibility of the refinement of the macrostructure of cast ingots of undoped high-purity molybdenum and tungsten for several times by increasing the melting rate is shown. Efficiency of the technological plasticity at the high-temperature treatment (pressing, rolling) of undoped polycrystalline ingots of molybdenum and tungsten is demonstrated. Unalloyed high-purity molybdenum has the recrystallization temperature of about 1000°C. The recrystallization annealing is used at different stages of the high-temperature processing the undoped high-purity molybdenum and tungsten. Mechanical testing shows that the undoped high-purity molybdenum has the optimistic characteristics of strength and plasticity in the temperature range from –190^{0}C up to 200^{0}C, which, at least, are not worse than the standard doped molybdenum alloys.

4. Of the structural factors facilitating brittleness in high-purity polycrystalline tungsten, it is turned out that the most significant factors are (1) the diffusion contamination of the subsurface layers with oxygen and hydrogen (by more than 1×10^{-4}% each with respect to the volume contents) during hot rolling and (2) the decreased density (less than 18.9 g cm^{-3}) of some tungsten ingots. Tungsten ingots of high purity are obtained using vacuum melting techniques. The distribution of pores and carbides in these ingots is investigated. It is found that the cast tungsten ingots with a density of less than 18.9 g cm^{-3} (the "low" density ingots), the sheets produced from these ingots can be fractured under hot rolling. Pores inherited from the initial cast structure facilitate cracking at cooling on the rolls during rolling. It is found that oxygen (more than 10^{-4}%) has a negative influence on tungsten during hot rolling because it has became extremely brittle. At the temperatures lower than the brittle–ductile transition temperature the sheets are fractured. In tungsten of the "low" density, hydrogen is accumulated in pores and cracks while ingots are cooling. This may facilitate brittle fracture of the sheets upon rolling. Within the range of deformation temperatures considered in this study, it is impossible to eliminate the pores inherited from the initial tungsten ingot of the "low" density.

Acknowledgements

I thank my colleagues and friends from the Institute of Solid State Physics, Chernogolovka, Russia, for their cooperation in my scientific research as well as in my life. My special gratitude to Hidde Brongersma and his excellent "Calipso" company. I am still very glad and proud that I spent many happy and productive years with him working in his scientific group at the Faculty of Physics at Technical University of Eindhoven, Eindhoven, Nederlands. I express my gratitude to the Russian Fund for Basic Research for financial support of the most part of these studies.

http://dx.doi.org/10.5772/59397

Author details

Vadim Glebovsky*

Address all correspondence to: glebovs@issp.ac.ru

Institute of Solid State Physics, the Russian Academy of Sciences, Russia

References

[1] Fromm E., Gebhardt E. Gase und Kohlenstoff in Metallen. Berlin. 1976.

[2] Blakely J.M., Shelton J.C. Equilibrium Adsorption and Segregation. In: Surface Physics of Materials, Academic Press; 1975. p100–107.

[3] Seah M.P., Lea C. Surface Segregation and its Relation to Grain Boundary Segregation. Philosophical Magazine 1975;31(3) 627–645.

[4] Lagues M., Domange J.L. Surface Segregation. Comparison between Theory and Experiment. Surface Science 1975; 47 77–85.

[5] Hoffmann S., Erlewein J. A Model of the Kinetics and Equilibria of Surface Segregation in the Monolayer Regime. Surface Science 1978;77 591–602.

[6] Farrel H.H., Isaacs H.S., Strongin M. The Interaction of Oxygen and Nitrogen with the Niobium (100) Surface. II. Reaction Kinetics. Surface Science 1973; 38 31–52.

[7] Pacia H., Dumesic J.A., Weber B., Cassuto A. High Temperature Low Pressure Reactions of Oxygen with Tantalum as Studied by AES and Line-of-sight Mass-spectrometry. Journal of Chemical Society Faraday Transactions 1976;9 1919–1934.

[8] Jupille J., Michel J.M. Desorption d'Oxydes de Niobium a partir de la Solution Solide Niobium-Oxygene, sous 3.10 torr a 3.10 torr d'Oxygen entre 1700 K et 2350 K. Journal of Less-Common Metals 1975;39(1) 17–34.

[9] Gomer R. Chemisorption on Metals. In: Solid State Physics, v.30, New-York-San-Francisco-London: Academic Press; 1975. p94–225.

[10] Muscat J.P., Rewns D.H. Chemisorption on Metals. Progress of Surface Science 1978;9(1) 1–43.

[11] Adamson A.W. Physical Chemistry of Surfaces. 3rd edition, New York: Wiley Interscience. 1980.

[12] Yates J., Madey T. Chemisorption on Rhenium: Nb and CO. Journal of Chemical Physics 1969;51(1) 339–397.

[13] Menzel D. Desorption Phenomena in Topics in Applied Physics. In: Gomer R. (ed.) Interaction on Metal Surfaces. Berlin-Heidelberg-New-York: Springer Verlag. 1975. p102–138.

[14] Weber B., Cassuto A. Elements d'un Modele d'Interactions Oxegene sous Basse Pression - Metaux de Transition a Haute Temperature. Surface Science 1973;39 83–99.

[15] Philippard J.L., Weber B., Cassuto A. Segregation Superficille de Rhenium Induite par l'Oxegene a Travers des Couches Mineces de Platine. Applied Surface Science 1982;10(1) 21–26.

[16] Doyen J., Ertl J. Theory of Carbon Monoxide Chemisorption on Transition Metals. Surface Science 1976;43(1) 197–229.

[17] Fromm E., Gebhardt E. Gase und Kohlenstoff in Metallen. Moscow: Metallurgy, 1980.

[18] Ono K., Moriyama J. Deoxidation of High-melting-point Metals and Alloys in Vacuum. Metallurgical Transactions B 1982;13(1–4) 241–249.

[19] Kulikov I.S. Deoxidation of Metals. Moscow: Metallurgy. 1975.

[20] Gillet E., Chiarena J.C., Gillet M. Decomposition Thermique du Monoxyde de Carbon sur une Surface de Molybdene: Formation de couches Superficielles de Carbon et de Carbure. Surface Science 1976;55 126–140.

[21] Ford R.R. Carbon Monoxide Adsorption on Transition Metals. In: Advances in Catalysis. New-York: Academic Press; 1970.

[22] Jayamatha K., Srinivasan A., Negbe M.S. Interactions of CO with Transition Metal Surfaces. Surface Science 1980;99(2) 309–319.

[23] Jupille J., Cassuto A. Interaction de l'Oxegene aux Basses Pressions avec la Solution Solid Niobium-Oxygene a Haute Temperature. I. Equilibre de Segregation-Chimisorption. Surface Science 1976;60 177–195.

[24] Goymor C.G., King D.A. Chemisorption of Carbon Monoxide on Tungsten. Journal of Chemical Society Faraday Transactions 1973;4 736–760.

[25] King D.A. Thermal Desorption from Metal Surfaces: a Review. Surface Science, 1975, v.47, p.384–402.

[26] Felter T.E., Estrup P.J. Adsorption of Carbon Monoxide on the Molybdenum (100) Surface. Surface Science 1978;76 464–482.

[27] Guillot G., Riwan R., Lecante J. Dissociation of CO on Mo(100). Surface Science 1976;59 581–592.

[28] Klein R., Leder L.B. Field-emission of CO on Tantalum. Journal of Chemical Physics 1963;38(8) 1866–1872.

[29] Gasser R.P.H., Thwaites R. Interaction of Carbon Monoxide with Tantalum. Transactions of Faraday Society 1965;65 2036–2043.

[30] Belov V.D., Ustinov Y.K., Komar A.P. Carbon Monoxide and Carbon Dioxide Interaction with Tantalum. Surface Science 1978;72 390–404.

[31] Ko S.M., Schmidt L.D. Adsorption and Solution of H_2, D_2, N_2, 0_2 and CO by (100)Ta. Surface Science 1975;47 557–568.

[32] Ustinov Y.K. On Determination of Parameters of Diffusion of Adsorbate in Metal Adsorbent by Flashing. Journal of Technical Physics 1972;42 2153–2157.

[33] Kosina L.N., Revyakin A.V., Samarin A.M. Thermodynamic Analysis of Deoxidation of Liquid Molybdenum. Reports of AN 1970;190(4) 909–910.

[34] Kosina L.N., Revyakin A.V., Samarin A.M. About Solubility of Oxygen in Liquid Molybdenum. Russian Metallurgy (Metalli) 1970;1 56–62.

[35] Kosina L.N. Deoxidation Ability of Carbon in Liquid Molybdenum. Russian Metallurgy (Metalli) 1970;2 116–118.

[36] Revyakin A.V., Kuznetsov L.V. Kinetics of Interaction of carbon and Oxygen in Liquid Metals in Vacuum. Russian Metallurgy (Metall) 1973;1 23–29.

[37] Kosina L.N., Revyakin A.V. Analysis of Deoxidation of Liquid Niobium. Russian Metallurgy (Metall) 1975;4 24–26.

[38] Efimov V.E. Kinetics and Mechanism of Interaction of Carbon and Oxygen in Liquid Refractory Metals. Russian Metallurgy (Metalli) 1975;4 27–34.

[39] Efimov V.E. Mechanisms of Interaction and Limiting Concentrations of Interstitials in Molybdenum. Russian Metallurgy (Metalli) 1977;1 39–45.

[40] Efimov V.E. About Interaction of Interstitialas with Metals during Vacuum Refining. Russian Metallurgy (Metalli) 1983;6 38–46.

[41] Efimov V.E. On the Kinetic and Mechanisms of Vacuum Refining Metals of Interstitials. In: Burkhanov G.S. (ed.); Physical and Chemical Basis of Vacuum Processing in Metallurgy. Moscow: Nauka; 1984. p36–42.

[42] Schulze K.K. Preparation and Characterization of Ultra-high Purity Niobium. Journal of Metals 1981;33(5) 33–41.

[43] Reed R.E. Electron-beam Float Zone Melting and Vacuum Degassing of Niobium Single Crystals. Journal of Vacuum Science & Technology 1972;9(6) 1413–1418.

[44] Meyerhogg R.W. Preparation and Electrical Resistivity of Ultrahigh Purity Niobium. Journal of Electrochemical Society 1971;118(6) 997–1001.

[45] Reed R.E. Redistribution of Ta and W Impurities in Niobium by Electron Beam Zone Melting. Journal of Crystal Growth 1973;19 61–67.

[46] Shipilevsky B.M., Glebovsky V.G. The interaction of carbon and oxygen in refractory metals at high temperatures. In: International Conference on High Purity Materials in Science and Technology. Dresden. 1980.

[47] Shipilevsky B.M., Glebovsky V.G. Competition of Bulk And Surface Processes in the Kinetics of Hydrogen and Nitrogen Evolution from Metals into Vacuum. Surface Science 1989;216 509–527.

[48] Shipilevsky B.M., Glebovsky V.G. The Laws of Interaction of Carbon and Oxygen Impurities at High Temperature Annealing of Refractory Metals in Vacuum. Poverkhnost (Surface) 1982;7 26–37.

[49] Shipilevsky B.M., Glebovsky V.G. The Features of Carbon Behavior in Refractory Metals at High-Temperature Annealing in Vacuum. Poverkhnost (Surface) 1982;8 115–119.

[50] Shipilevsky B.M., Glebovsky V.G. Characteristic Features of Interaction between Carbon and Oxygen Impurities during High-temperature Annealing of Refractory Metals in Vacuum. Physics, Chemistry & Mechanics of Surfaces 1982;1(7) 1940–1964.

[51] Shipilevsky B.M., Glebovsky V.G. Features of Carbon Behavior during High-temperature Annealing of Refractory Metals in Vacuum. Physics, Chemistry & Mechanics of Surfaces 1982;1(8) 2291–2310.

[52] Shipilevsky B.M., Glebovsky V.G. Competition of the Bulk and Surface Processes in the Kinetics of Hydrogen and Nitrogen Release from Metals into Vacuum. Vacuum 1990;41 126–129.

[53] Glebovsky V.G., Kapchenko I.V., Shipilevsky B.M. Kinetics of Interaction of Carbon and Oxygen in Niobium during Vacuum Levitation Melting. Journal of Alloys and Compounds 1992;184 305–313.

[54] Kapchenko I.V., Glebovsky V.G., Kireyko V.V., Ryzhkov V.A. Kinetics of Removal of Oxygen and Carbon from Molybdenum during High-frequency Levitation in Vacuum. Vysokochistye Veschestva (High-Purity Materials) 1989; 114–124.

[55] Glebovsky V.G., Shipilevsky B.M., Kapchenko I.V., Kyreiko V.V. Kinetics of Oxygen and Carbon Removal from Liquid Molybdenum in the Process of High-frequency Levitation in Vacuum. Journal of Alloys and Compounds 1992;184 297–304.

[56] Aritomi N. Research on the Brittleness of Electron-Beam Melted Molybdenum. Transactions of Japan National Research Institute of Metals 1979;21 18–34.

[57] Morito F. Intergranular Fracture Surface Analysis of Molybdenum. Surface & Interface Analysis 1990;15 427–432.

[58] Kishore R., Kumar A. Effect of Carbon on Ductilization of Electron-beam Welds in Molybdenum. Journal of Nuclear Materials 1981;101 16–27.

http://dx.doi.org/10.5772/59397

[59] Glebovsky V.G., Kasyanov V.G., Kapchenko I.V., Podvigin V.S. Macrostructure and Plasticity of Molybdenum. High Temperatures-High Pressures 1986;18 453–458.

[60] Glebovsky V.G., Karpov M.I., Kasyanov V.G., Kokhanchik L.S., Medved N.V. Technological Plasticity of Unalloyed Polycrystalline EB-melted Molybdenum. T-00615. Chernogolovka: Institute of Solid State Physics; 1985.

[61] Glebovsky V.G., Kapchenko I.V., Kosachev L.S., Leitman M.S., Podvigin V.N. Effect of some Melting Parameters on Structure and Properties of Unalloyed Molybdenum. In: Burkhanov G.S. (ed.), Study of Refractory Materials: Proceedings of the National Conference for Refractory Metals. Moscow: Nauka Publishers; 1988.

[62] Glebovsky V.G., Karpov M.I., Kasyanov V.G., Kokhanchik L.S., Medved N.V. Macrostructure and Technological Ductility of Unalloyed Molybdenum Produced by Electron Beam Melting. Russian Metallurgy (Metalli) 1986; 112–116.

[63] Gnesin B.A., Glebovsky V.G., Karpov M.I., Kireyko V.V., Snegirev A.A. Some Structural Sources of Brittleness in Cast Tungsten. Journal of Less-Common Metals 1990;167 11–19.

[64] Gnesin B.A., Glebovsky V.G., Karpov M.I., Kireyko V.V., Snegirev A.A. Structural Sources of Brittleness of High-purity Polycrystalline Tungsten. Vysokochistye Veschestva (High-Purity Materials) 1991; 199–203.

[65] Glebovsky V.G., Kapchenko I.V., Kosachev L.S., Egorkin M.I. Structural Sources of Fragility of Tungsten Melted in Vacuum. Burkhanov G.S. (ed.), Proceedings of National Conference on Refractory Metals. Suzdal: Informelectro. 1987.

[66] Medvedev G. A., Denisov P. I., Medvedev A. G. The Method of Metal Temperature Evaluation at Hot Rolling of Sheets and Strips. Sverdlovsk: UPI, 1981.

[67] Kopecky C.V. Structure and Properties of Refractory Metals. Moscow: Metallurgy, 1974.

[68] Novikov I.I., Klimov K.M, Burkhanov Yu.S. About Estimation of Temperature at Hot Rolling. Russian Metallurgy (Metalli) 1983;2 88–92.

[69] Gmelins Handbuch der Anorganischen Chemie, N54 W; Teil Bl. Berlin: Springer Verlag. 1979.

Chapter 3

Recrystallization of Rolled α-Zr Single Crystals

Yu Perlovich and M. Isaenkova

Additional information is available at the end of the chapter

http://dx.doi.org/10.5772/60019

1. Introduction

Recrystallization of metal material is a complicated phenomenon that unites different aspects of its profile, including plastic deformation, strain hardening and behaviour by heat treatment. The latter entails numerous processes such as recovery, growth of recrystallization nuclei and the competition of new grains. When considering the plastic deformation of single crystals, the experimental analysis of their ensuing recrystallization becomes simpler: in the grains of a polycrystal, plastic deformation creates a wide spectrum of substructures depending on the crystallographic orientation of grains, whereas in the case of a deformed single crystal, initial substructure conditions are much more limited.

In principle, concepts of 'polycrystal' and 'single crystal' ought to be associated with the concept of that of 'deformed single crystal'. The single crystal, subjected to plastic deformation (most often one deals with rolled single crystals) does not become a polycrystal; at the same time, it ceases to be a single crystal, because of the crystallographic texture that forms in it due to the number of initial orientations of this single crystal. A rolled polycrystal is a conglomerate of rolled single crystals, but a rolled single crystal differs from a polycrystal (or a rolled polycrystal) in terms of boundary characteristics, which separate regions with different orientations. In the polycrystal, these regions correspond to individual grains and are separated from neighbours by high-angular boundaries. In the rolled single crystal, these regions originate from the same monocrystalline matrix, due to the inevitable non-uniformity of plastic deformation; therefore, they are separated from neighbours by low-angular boundaries, except for the case of twinning, which was most often observed in the current work.

Experiments involving the rolling of single crystals can provide results that may be unreachable by similar experiments with polycrystals. Among such results is the obtaining of quasi-monocrystalline foils by the cold rolling of BCC single crystals (Mo, Nb) in the case of initial orientation {001}<011>. These foils are characterized by minimal strain hardening up to the highest deformation degrees and do not recrystallize by subsequent annealing [1]. At the same time, rolled single crystals with other initial orientations acquire significant strain hardening and can even be destroyed under rolling. These effects make clear the mechanisms of plastic deformation and recrystallization of materials, and indicate how the substructure of polycrystalline products can be modified and optimized. To date, the existence of analogous effects in the rolling of α-Zr single crystals have not been investigated due to difficulties related to procuring them.

Though the title of this chapter is "Recrystallization of rolled α-Zr single crystals", the intention of authors is wider and concerns also recrystallization of rolled polycrystalline α-Zr, since the behaviour under annealing of rolled single crystals helps to explain some known features of recrystallization in polycrystals and can compare the characteristics of strain hardening distribution in rolled single crystals and polycrystals.

2. Reorientation of grains and texture development in α-Zr under rolling

When considering recrystallization and especially its textures, it is necessary to take into account the prehistory of the material, including features of its previous deformation, grain reorientations, formed deformation textures and plastic deformation mechanisms. Therefore analysis of recrystallization inevitably expands in scope and covers a number of adjacent questions, among them, the details concerning preceding grain reorientation. Thus, in the current chapter, recrystallization is considered as a derivative effect of plastic deformation.

As a preamble to discussing the behaviour of α-Zr single crystals under rolling, we take into account the Hobson's calculated diagram for operating plastic deformation mechanisms [2] and our X-ray experimental observations of basal axes (or axis) trajectories during the course of rolling [3, 4]. The Hobson's diagram, added later by Matcegorin et al. [5] (Fig. 1), predicts slip and twinning systems operating in α-Zr grains in conditions of cold rolling. Three twinning systems are considered: (1) $\{10\bar{1}2\}\langle\bar{1}011\rangle$ and (2) $\{11\bar{2}1\}\langle\bar{1}\bar{1}26\rangle$, activated by the tension between angles of basal axes (or axis) reorientations at 85° and 35°, respectively; (3) $\{11\bar{2}2\}\langle\bar{1}\bar{1}23\rangle$, activated by compression and with a 64° angle of reorientation. In [2] and [5], for polycrystalline Zr-based alloy regions, the stereographic projections are drawn under which rolling prismatic, basal and pyramidal slip systems operate. Experimental trajectories of basal axes in cold-rolled α-Zr grains [3, 4, 6] contain jump-like segments with an angular length of 10-15°, which testify in particular to possible twinning by corresponding systems (Fig. 2-3). The unique evidence of twinning in the annealed polycrystalline plate of the Zr-1%Nb alloy consists in the sharp jump-like reorientation of basal axes in the surface layer of a plate at the first moment of its tensile deformation along a transverse direction. Due to twinning basal axes change their orientations in the jump-like manner and get from vicinity of pole figure diameter ND-TD to regions near its diameter ND-RD [4].

Additionally, it was found [6] that development of the rolling texture in polycrystalline α-Zr has three stages:

1. formation of the quasi-stable texture T1 (0001)±15-20°ND-RD $\langle 11\bar{2}L\rangle$

http://dx.doi.org/10.5772/60019

2. transition T1 → T2,

3. perfection of the stable texture T2 (0001)±30-40°ND-TD <10$\bar{1}$0>.

Texture T1 is supported due to an equilibrated slip at the basal and pyramidal planes, transition T1 → T2 is connected with the strain hardening of pyramidal slip systems and the stability of the final rolling texture T2 is conditioned by an equilibrated slip through basal and prismatic planes (Fig. 4). The jump-like reorientation of basal axes due to twinning by planes {11$\bar{2}$1}, revealed experimentally, throws them into the neighbouring region of the basal slip shown in the diagram in Fig. 1 [4], thereby moving basal axes further to the centre of stereographic projection. The continuous arrows coinciding with experimentally observed reorientation trajectories through the slip of basal axes (Fig. 4) are finally directed to the maxima of rolling textures T1 or T2. This indicates that in polycrystalline α-Zr, the rolling texture develops mainly by means of basal, pyramidal and prismatic slip.

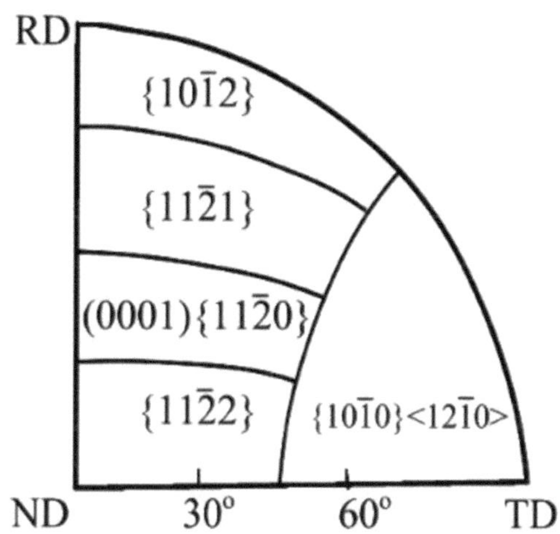

Figure 1. Regions of stereographic projection for the α-Zr sample, whereby its rolling, different deformation systems operate depending on the orientation of the grain basal axis. Calculated systems are characterized by the largest values of the Schmidt factor. Assumptions were made that each system has the same critical stress and that $\sigma_{RD} = \sigma_{TD}$[5].

Development of the rolling texture in α-Zr single crystals was studied as applied to coarse-grained samples [7], since lattice reorientation in separate coarse grains as a first approximation occurs independently of similar processes in other grains. However, the formation of the recrystallization texture in the same coarse-grained sample includes the growth of nuclei at grain boundaries, resulting in this process moving outside of separate grains and its control via the orientations of neighbouring grains. Therefore, recrystallization in rolled coarse-grained samples does not imitate the situation in rolled single crystals. Thus, the current study is the first X-ray investigation of recrystallization in cold-rolled α-Zr single crystals.

As for recrystallization of cold-rolled polycrystalline α-Zr, its regularities were described in detail in [8-9]: reorientation of prismatic axes by 30° about the basal axis, the shift of basal axes along the TD-ND diameter of stereographic projection in the direction of ND and in the case of multicomponent rolling, texture redistribution of its components according to definite relationships. However, these data were obtained only by the geometric analysis of texture features, without the use of data concerning the distribution of strain hardening in rolled polycrystals. Therefore, the motive power of recrystallization in the considered concrete cases is unclear. Currently, the X-ray method of generalized pole figures is being developed [10, 11], which allows revealing the distribution of strain hardening in textured materials, depending on the crystallographic orientation of their grains. This will allow a better understanding of the reasons for the observed passage of recrystallization.

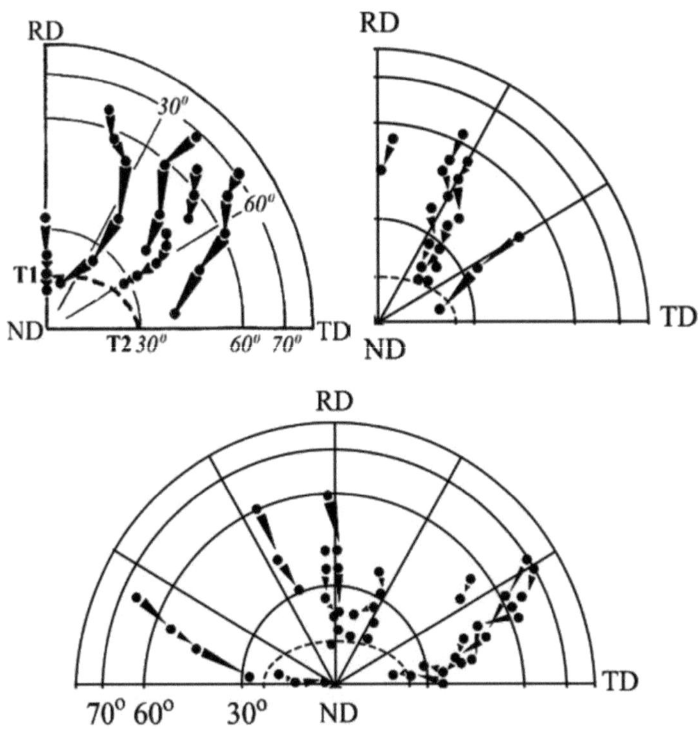

Figure 2. Examples of the experimental trajectories of basal axes reorientations during the course of cold rolling of Zr and its alloys.

http://dx.doi.org/10.5772/60019

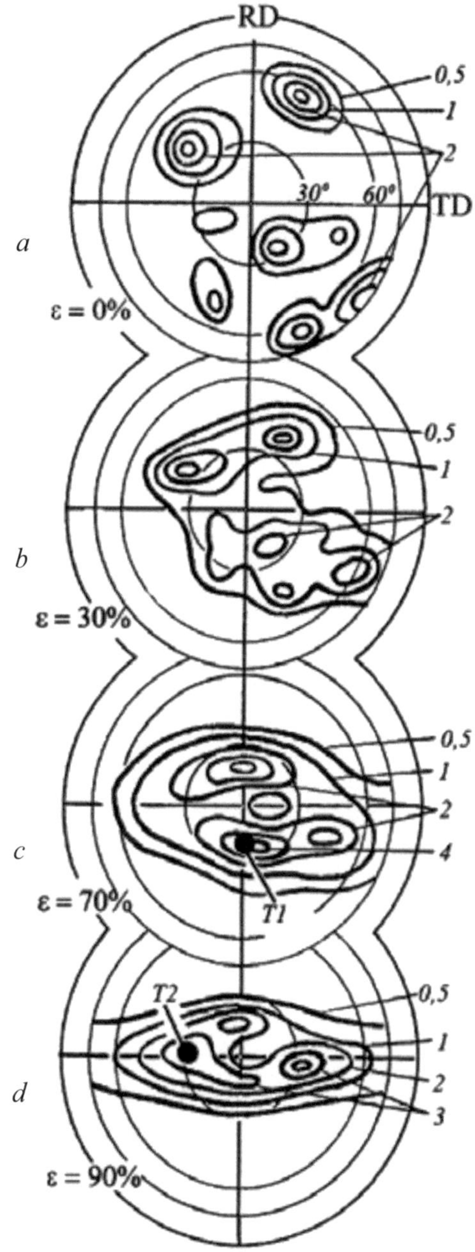

Figure 3. Evolution of PF (0001) for α-Zr by increasing the deformation degree through cold rolling.

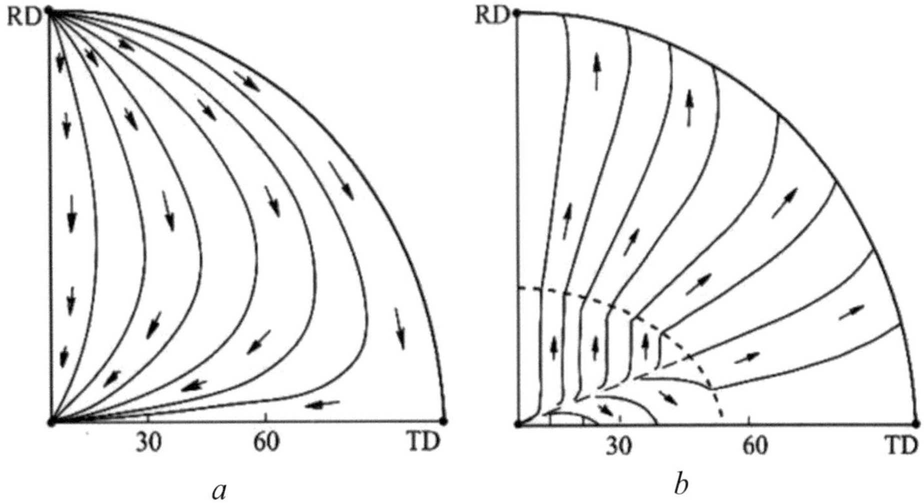

Figure 4. Calculated trajectories of basal axes reorientations during the course of rolling through the operation of basal (a) and pyramidal (b) slip systems [5].

3. Experimental procedure: Studied samples and used X-ray methods

A single crystal of pure Zr was obtained by means of solid-phase over-crystallization. Samples 7x10x3 mm in size were cut from the cylindrical single crystal using the electro-erosion method. Obtained samples were rolled at a laboratory rolling mill between plates of stainless steel up to the deformation degree ε_{max}= 80% and with ~5% reductions per pass.

The initial monocrystalline plate and rolled single crystals were subjected to X-ray texture investigation, which included (1) texture analysis using the method of direct pole figures; (2) the study of the strain hardening distribution using the method of generalized pole figures (GPF) (see the next section).

Then, rolled single crystals were annealed in the evacuated vessel at 580°C for 3 h, the standard regime of recrystallization heat treatment for industrial products made from Zr-based alloys. After annealing, rolled single crystals were repeatedly subjected to X-ray investigation; direct pole figures were measured once more and subtraction diagrams were constructed on the basis of texture data treatment (see the next section).

In addition to the investigation of rolled α-Zr single crystals, the current work also included the study of cold-rolled and recrystallized (1) coarse-grained iodide Zr and (2) a polycrystalline sheet of Zr-1%Nb alloy, rolled along RD and along TD. The development of rolling textures, distributions of strain hardening and recrystallization textures in these samples was studied using the same X-ray methods as those used in the study of rolled single crystals. When comparing X-ray data obtained for rolled single crystals and polycrystalline samples, a number of important conclusions concerning the regularities of their strain hardening and recrystallization were made.

All X-ray studies of the used samples were carried out using an X-ray diffractometer Bruker D8 DISCOVER with a LynxEye position-sensitive detector. Partial direct pole figures PF(0001) and {11$\bar{2}$0} were constructed with an angular radius 80° [12].

3.1. The method of Generalized Pole Figures (GPF)

For rolled α-Zr single crystals, the X-ray method of generalized pole figures (GPF) was used. This method consists of registration of X-ray line profiles during the course of texture measurement [10, 11]. In the current case, the method characterized the condition of the crystalline lattice along basal axes <0001> in all grains of the sample. The physical angular half-width β_{0002} of the X-ray line (0002), increasing as a result of strain hardening and decreasing by recrystallization was used as a measure of crystalline lattice distortion. A wide spectrum of substructure conditions, i.e., values of β_{0002}, the distribution of which in the stereographic projection is shown with GPF β_{0002}, corresponded to separate micro-fragments of the rolled single crystal or grains of rolled polycrystal, and the optimal criterion for systematization of GPF data was their crystallographic orientation.

In particular, for many rolled metal materials with the developed texture it was revealed that the strain hardening of fragments (grains, subgrains, crystallites) corresponding to texture maxima was significantly lower than that for texture minima. Strain hardening increased as grain orientation shifted from a texture maximum to a texture minimum. In cold-rolled BCC metals, where primary recrystallization is realized by the growth of grains with increased strain hardening, new maxima of the recrystallization texture grew at the maxima "slopes" of the rolling texture. Hence, strain hardening of these grains was sufficiently high for their predominant growth; at the same time, there were a sufficient amount of grains with such orientation for them to be able to "swallow" the rest of the deformed matrix.

Application of the GPF method requires the availability of an X-ray texture diffractometer of last models with a position sensitive detector, which will allow registering the entire X-ray line profile by texture measurement, without time losses for movement of the detector. The procedure of GPF measurement and construction requires special software, which can be developed on the basis of the standard technical potential of BRUKER diffractometers.

3.2. The method of mutual subtraction of texture Pole Figures (PF)

In order to reveal regions of pole density changes in PF(0001) through the transition from cold-rolled to recrystallized α-Zr, difference diagrams $S(\psi,\varphi)$ are constructed in which (ψ,φ) are the angular coordinates of points in PF and $S(\psi, \varphi)$ is equal to the difference between pole densities in corresponding points of PF(0001) for recrystallized and cold-rolled samples; thus, $S(\psi, \varphi) = [PF(0001)_{recr}\text{-}PF(0001)_{roll}]$. Inhomogeneous strain hardening and recovery result in some violation of proportionality between scattering ability, calculated pole density and specific volumes of deformed and recrystallized grains. In this connection, pole densities in the PF of rolled and recrystallized samples can be compared only with limited accuracy. Therefore, subtraction diagrams are constructed only for those regions of PF(0001)$_{roll}$ where pole densities noticeably exceed the sensitivity of texture measurements. In the current case, this threshold value of pole density was accepted as 0.5. Represented below, subtraction diagrams visually demonstrate the arrangement of regions with maximal changes of pole density as a result of recrystallization.

4. Experimental results

4.1. Twinning as the main deformation mechanism of α-Zr single crystals

In Fig. 5a, the partial direct PF{11$\bar{2}$0} with an angular radius of 80° is presented for the initial α-Zr single crystal. The axis of the initial cylindrical single crystal is situated in the centre of PF; it coincides with the normal direction of the plate cut from the crystal and is perpendicular to the direction of its following rolling. In the same PF, near the circumference of stereographic projection with an angular radius of 80° the exit of the basal axis [0001] is shown. This is represented by concentric circular contours and is distanced by an angle of 90° from the exits of prismatic axes <11$\bar{2}$0>. After rolling of the single crystal up to a deformation degree of ~5%, essential changes take place in its stereographic projection: jumping from the PF(0001) shown in Fig. 5b, due to the activation of twinning by planes {11$\bar{2}$1}, the basal axis reorients by ~35°, whereas due to twinning by plains {10$\bar{1}$2} reorientation occurs by ~85°. When rolling the single crystal by 53%, the texture maximum formed by twinning strengthens due to further twinning in regions of the single crystal with the initial orientation and additionally shifts by ~15°-20° due to basal slip [5] (Fig. 5-*c*). The ensuing rolling of up to ε=80% leads to formation in the rolled single crystal of a two-component texture (Fig. 5d) typical for α-Zr [6, 13].

The same cycle of investigations was repeated for monocrystalline plates of three initial orientations, differing only in terms of RD. These orientations corresponded to the following angular coordinates of basal axes: N 1 - (70°, 33°); N 2 - (70°, 152°); N 3 - (70°, 87°). The three columns of PFs in Fig. 5 demonstrate the successive stages of rolling texture development in single crystals within the following indicated orientations: N 1 - (*a-d*), N 2 - (*e-i*), N 3 - (*j-n*). The first stages of all three successions in Fig. 5 corresponded to the deformation degree ε≈5-7% in order to create mutually differing distributions of monocrystalline fragments resulting from various positions of twinning systems relative to the RD. By considering the obtained data it became evident that twinning was the main plastic deformation mechanism of α-Zr single crystals.

After deformation by only 5-7%, the texture maximum in PF(0001) characterizing the initial orientation of a single crystal broke into five-six new maxima corresponding to the reorientation of single crystal fragments by means of twinning in planes {10$\bar{1}$2}, {10$\bar{1}$1} and {11$\bar{2}$1} (Fig. 5b, e, j). The initial orientation of prismatic axes of the single crystal (Fig. 5a) and the characteristic reorientation angles of basal axes at 85°, 57° or 123° and 35°, shown in Fig. 5b, e, j, allowed for identifying unique operating systems of twinning. The primary planes of twinning in these cases were {10$\bar{1}$2} and {11$\bar{2}$1}, resulting in the formation of texture maxima, which are indicated by red and blue arrows, respectively (see Fig. 5-b, e, j). Consequently, maxima situated near ND had the greatest intensity. Additional weak texture maxima, marked by Figures 1-4 in Fig. 5e and j, formed due to twinning of the initial single crystal by planes {10$\bar{1}$1} (green arrows in Fig. 5e). These maxima proved to be unstable due to their orientation (Fig. 5f) and disappeared at subsequent stages of rolling, providing texture scattering (Fig. 5g). Twinning by planes {10$\bar{1}$1} had previously not been considered in the analysis of rolling texture formation in α-Zr [2, 5-7].

The above-described changes of basal axes orientations when rolling α-Zr single crystals with orientations N 1 and N 2 agreed with the proposed diagram by Hobson of plastic deformation mechanisms activated in α-Zr grains with different orientations of basal axes relative to external directions ND, RD and TD [2, 5]. The process of twinning continues until where some regions with the initial orientation remain in the single crystal.

This process is described using curves that show changes in PF parameters as the deformation degree increases (Fig. 6). Among these parameters, there are angular positions ψ and intensities P of initial texture maxima *A* and new texture maxima *B*, formed by twinning. Solid lines in Fig. 6 show changes in position and intensity of initial maxima *A*, whereas dotted lines indicate changes of position and intensity of derived maxima *B* for rolled single crystals N 2 (*a, b*) and N 3 (*c, d*). In particular, one can see that in both single crystals, derivative maxima *B* arose at deformation degrees 3-5%, whereas maximum *A* in PF(0001) for the single crystal N 2 disappeared by 13% after rolling. The decrease in P intensity for all maxima indicates the further fragmentation of the single crystal by means of twinning, including regions corresponding to both *A* and *B* maxima. The shift of maximum *A* in PF(0001) for the single crystal N 3 was connected with operation of the slip. In this single crystal, the basal axis of which was oriented near TD, twinning by planes {10$\bar{1}$2} also activated.

At the next stages of deformation in the single crystal N 2, twinning by planes {11$\bar{2}$1} was found. In Fig. 5f, along with initially formed texture maxima, new maxima arose, conditioned by secondary twinning in grains with orientation *B* by planes {11$\bar{2}$1}. Twinning by these planes also continued up to the exhaustion of grains with an orientation corresponding to texture T1. The possibility of twinning explains the transition of texture T1 into texture T2, which consisted of a shift in texture maxima from the diameter ND-RD of stereographic projection to their final stable position at the diameter ND-TD. Thus, along with pyramidal, basal and prismatic slip caused by deformation degrees exceeding 50%, twinning by planes {11$\bar{2}$1} also proved to be active in α-Zr. This fact did not agree with the conclusions of [14] concerning the suppression of twinning through the increase of deformation degree due to distortion of the deformed crystalline lattice.

Thus, up to ε=30% texture formation under rolling was provided by the active operation of twinning by planes {10$\bar{1}$2}, {10$\bar{1}$1}, {11$\bar{2}$1} and {11$\bar{2}$2}. When the deformation degree exceeded 30%, twinning by planes {11$\bar{2}$1} remained the active plastic deformation mechanism.

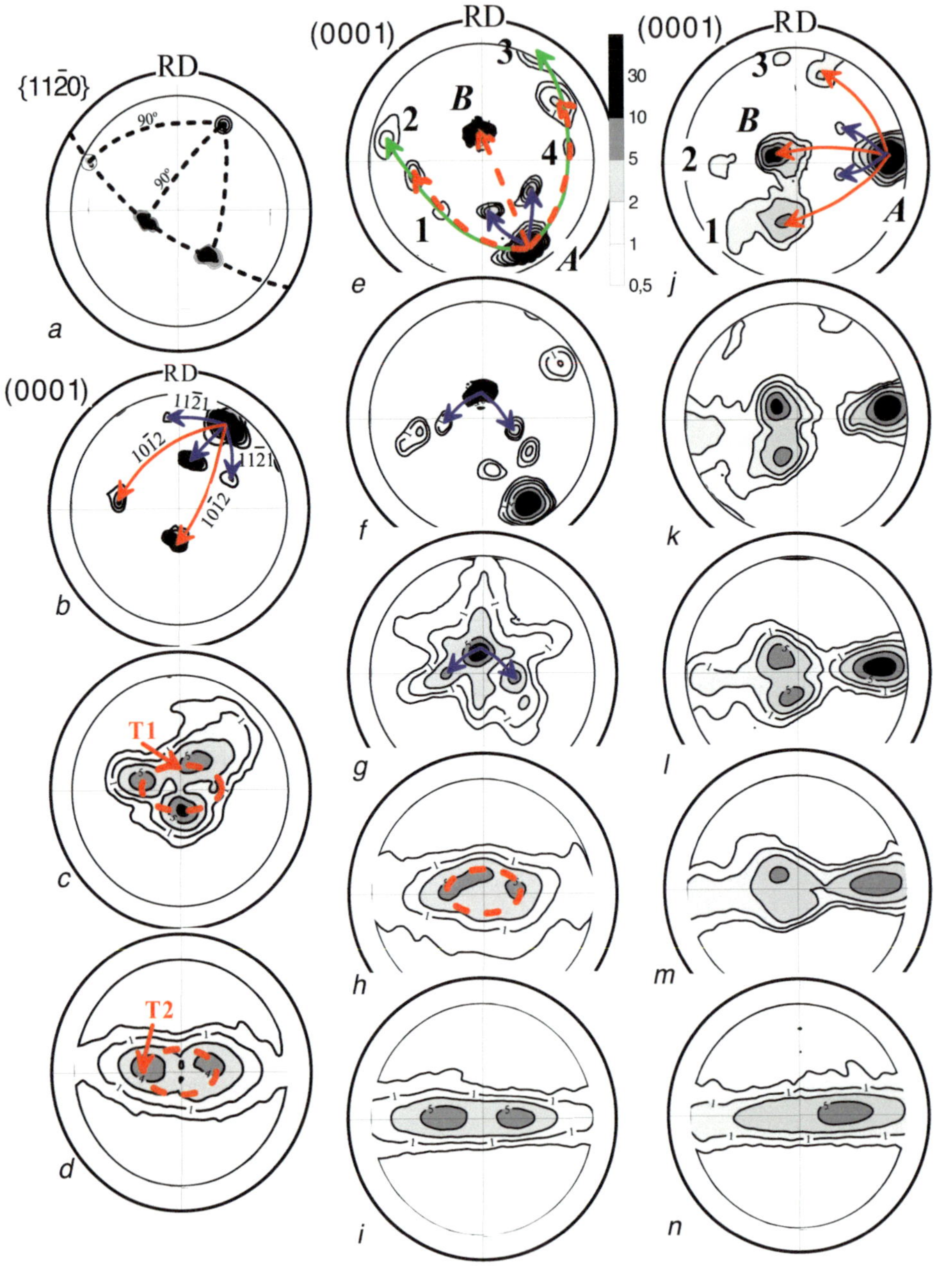

Single crystal N1:*b)*ε≈5 %;*c)*ε≈53 %;*d)* –ε≈80 %;
Single crystal N2: *e)*ε≈5 %;*f)*ε≈13 %; *g)*ε≈51 %; *h)*ε≈71 %; *i)*ε≈90 %;
Single crystal N3:*j)*ε≈7 %;*k)*ε≈30 %; *l)*ε≈50 %; *m)*ε≈70 %;*n)*ε≈98 %

Figure 5. PF{11$\bar{2}$0} of initial zirconium single crystal (a) and changes of PF (0001)due to twinning upon the rolling of α-Zr single crystals with different orientations:

4.2. Texture changes by slip

Fig. 5b and c, and Fig. 6c show that twinned areas of single crystals (fragments separated from other parts of a single crystal by planes of twinning) shift gradually to the centre of PF(0001), i.e., to ND of the rolled plate. Such a shift of texture maxima is conditioned by the development of a basal slip in the material. Along with clearly expressed additional texture maxima arising (see Fig. 5j), situated at the diameter ND-TD and symmetrically relative to ND, and connected with twinning by planes {11$\bar{2}$1}, formation of the elliptical texture maxima around ND can be observed. This maximum is markedby the dotted line in Fig. 5c and h, and serves as a boundary between regions of stereographic projection, with predominant activation of the pyramidal and basal slip systems [5], which provides the stability of the observed texture by intermediate deformation degrees of 50-70%. Observed gradual changes of the texture maximum position in Fig. 5k-n and Fig. 6c can be connected only with operation of the basal slip or through twinning by planes {11$\bar{2}$1}, resulting in the asymmetry of PF(0001) for rolled single crystal N 3, even at a deformation degree of 98%.

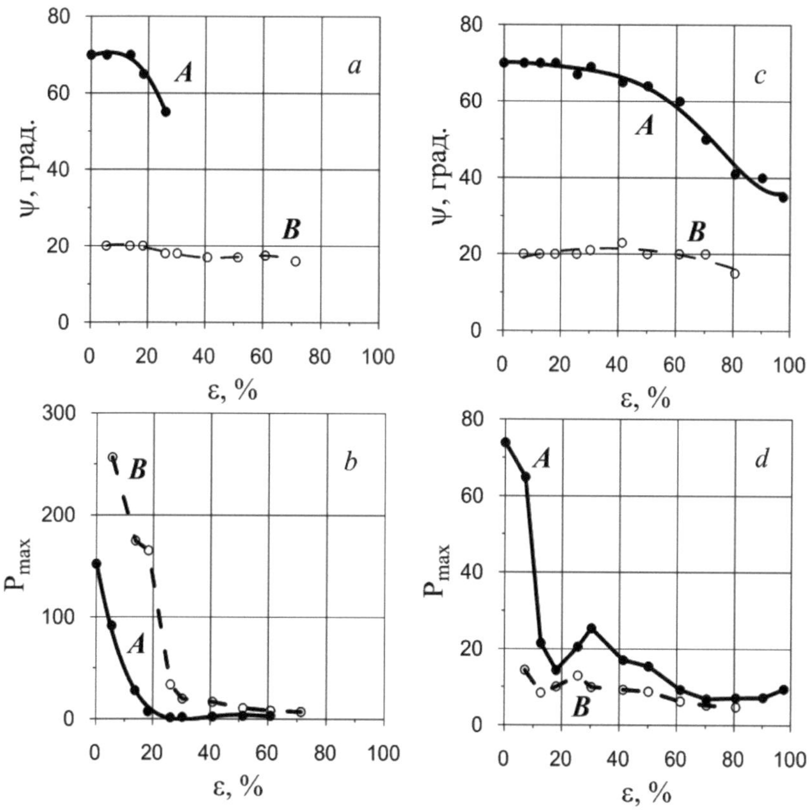

Figure 6. Changes in the pole coordinate ψ (a, c) and intensity (b, d) of texture maxima *A* and *B* (see Figure 5i, j) by increasing the deformation degree of α-Zr single crystals N 2 (a, b) and N 3 (c, d).

4.3. Recrystallization of rolled α-Zr single crystals

Results of the X-ray study of recrystallization in rolled α-Zr single crystals are presented in Fig. 7 and 8 for single crystal N 1, in Fig. 9 and 10 for single crystal N 2 and in Fig. 11 and 12 for single crystal N 3. The texture PF(0001) for rolled samples and recrystallized samples are shown in these figures as (*a*) and (*b*), respectively. For single crystals N 1 rolled at 53% and 80%, GPF β_{0002} were measured (Figs. 7c and 8c). For all rolled and recrystallized single crystals, subtraction diagrams $S(\psi, \varphi) = [PF(0001)_{recr} - PF(0001)_{roll}]$ are shown as (*d*). In Figs. 7e and 8e, diagrams of correlation between PF(0001) and GPF β_{0002} are presented. Each point in the diagrams corresponds to points with coordinates (ψ, φ) in PF and GPF at that values of pole density $P(\psi, \varphi)$, and are indicated in the axis of abscissas, whereas values of X-ray line half-width β_{0002} are showing the axis of ordinates.

The X-ray method of GPF, used for rolled single crystals in the current case characterized the condition of the crystalline lattice along basal axes <0001> in all fragments of the sample. As a measure of lattice distortion, which strengthens as a result of strain hardening and is removed by recrystallization, the physical angular half-width of the X-ray line (0002) β_{0002} was applied. A wide spectrum of substructure conditions, i.e., values of β_{0002}, responded to separate fragments of the rolled single crystal. In subtraction diagrams $S(\psi, \varphi) = [PF(0001)_{recr} - PF(0001)_{roll}]$ for each point (ψ, φ) of the stereographic projection, the difference between the pole densities of recrystallized and a rolled single crystal was calculated. The presented subtraction diagrams visually demonstrated the arrangements of regions where maximal changes of pole density by recrystallization were observed. Gradations of red colour in the subtraction diagram indicate regions of positive values and respond to an increase in pole density as a result of recrystallization, whereas gradations in blue colour correspond to a decrease of pole density due to recrystallization. Diagrams of correlation between PF(0001) and GPF β_{0002}, shown in Figs. 7e and 8e, had an appearance typical to that of a first approximation for many deformed metal materials, characterized by a gradual decrease of values β_{0002} as the peak of the texture maximum approached. Against a background of this tendency features can be observed as clearly connected to the character of corresponding rolling textures.

Due to the use of rolled α-Zr single crystals, the rolling textures obtained were formed mainly by twinning. In particular, in a single crystal N 1 rolled at 53%, i.e., by means of twinning, the rolling texture arose with maxima, which in spite of a relatively high deformation degree, took up transition positions, distinguishing them from maxima in stable textures T1 and T2. Subtraction diagrams showed that in the case of recrystallization within regions of PF for a rolled single crystal, where texture maxima arose due to twinning, the growth of pole density was absent (Fig. 7a, d). In fragments of rolled single crystals corresponding to the orientation of their basal axes near normal direction ND, rolling can activate only a pyramidal slip, characterized by high values of critical shear stress. Operation of other slip systems in these fragments (or grains of polycrystal) is suppressed due to low Schmidt factor values. Therefore, in fragments of α-Zr single crystals (or the grains of a polycrystal) experiencing reorientation, predominantly through the participation of twinning, their strain hardening, including distortion of the crystalline lattice and fragmentation of substructure elements, is relatively small. So, during the course of subsequent annealing, nuclei of recrystallization do not arise in these regions and these fragments (or grains of polycrystal) are swallowed up by recrystallization nuclei, growing more actively and corresponding by their orientation to regions of PF with increased strain hardening. GPF β_{0002} for rolled single crystals N 1 in Figs. 7c and 8c also indicate that the strain hardening of fragments with a basal axis orientation near ND is minimal. For this reason, in subtraction diagrams constructed for all rolled single crystals, we see a blue centre, testifying that in the central region the growth of pole density by recrystallization was absent, since the only mechanism provided here in terms of its input was twinning.

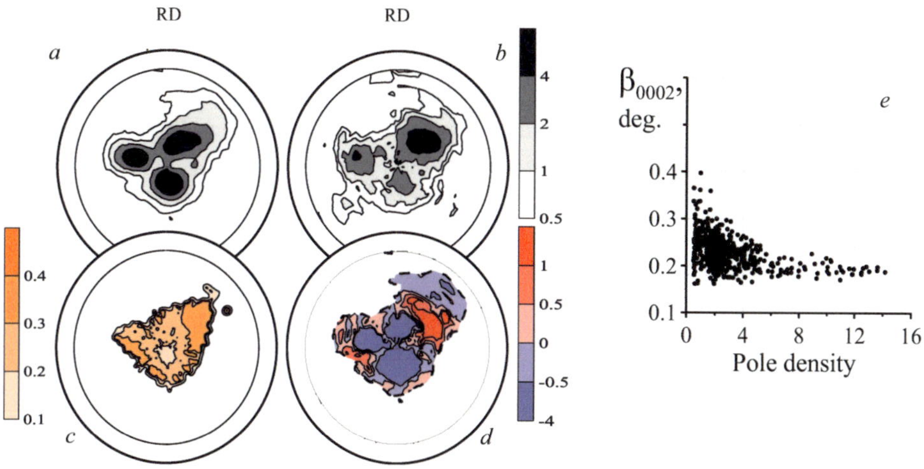

Figure 7. Recrystallization in cold-rolled α-Zr single crystal N 1 (ε~50%): *a* – PF(0001)$_{roll}$, *b* - PF(0001)$_{recr}$; *c* – GPF β_{0002}; *d* – [PF(0001)$_{recr}$- PF(0001)$_{roll}$]; *e* – diagram of correlations between PF(0001) and GPFβ_{0002}.

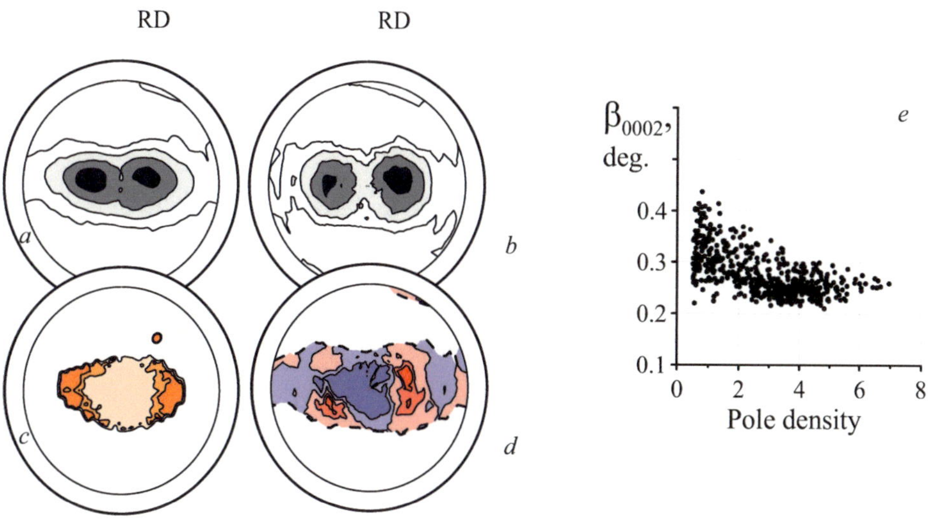

Figure 8. Recrystallization in cold-rolled α-Zr single crystal N 1 (ε~80%): *a* – PF(0001)$_{roll}$, *b* - PF(0001)$_{recr}$; *c* – GPF β_{0002}; *d* – [PF(0001)$_{recr}$- PF(0001)$_{roll}$]; *e* – diagram of correlation between PF(0001) and GPFβ_{0002}.

At the same time, the subtraction diagrams constructed for rolled single crystals show that its texture maxima are bordered by zones of intensive recrystallization (Fig. 7d). Only within these regions of stereographic projection was the increased lattice distortion also observed (Fig. 7c) as a consequence of the above-described simultaneous operation of the pyramidal and basal slip. According to GPF β_{0002}, the "slopes" of texture maxima (see Fig. 7c) respond to regions of greatest fragmentation and the distortion of the crystalline structure, so that in the current case, the main mechanism of α-Zr recrystallization was the intensive growth of regions in the deformed crystalline matrix with the increased energy of residual distortions.

As the deformation degree of single crystals increased, their rolling texture changed and acquired features of the texture T2 (Fig. 7-12). This testified that the relative input of twinning in the formation of new maxima gradually decreased and was replaced by the slip. When this happened, the character of subtraction diagrams and diagrams of correlation PF(0001) – GPF β_{0002} sharply changed. In the subtraction diagrams, additional red regions arose, coinciding approximately with the positions of T2 maxima in rolling textures. Hence, the strain hardening at these maxima was higher than in regions where twinning had previously been predominant and here, a greater number of recrystallization nuclei arose (Fig. 8, 10, 12).

The most detailed and precise data concerning the evolution of strain hardening distributions in rolled single crystals provided diagrams of correlation PF(0001) – GPF β_{0002}, showing in Figs. 7e and 8e. Texture maxima, formed by twinning, were twice as high as those formed by the slip at a later stage. Strain hardening of fragments corresponding to these maxima by ε=53% was constant and remained low (β_{0002}< 0,2°) for the most part, whereas for ε=80% strain hardening as noticeably higher (β_{0002}> 0,25°) and varied within significant intervals. The correlation diagrams show that strain hardening grew as an angular distance of the considered fragment from the top of maxima increases. The reason for this regularity was the different features of the slip within the stable texture maxima and at their periphery parts, i.e. at texture minima, where pole density is lower than 1. Stable texture maxima arose in regions of stereographic projection, where the operation of different slip systems was equilibrated. In the case of α-Zr rolling stability of T2, texture maxima were conditioned through the mutually balanced operation of basal and prismatic slip systems. However, the joint activity of these slip systems resulted in increased strain hardening, whereas their alternating operation, realized in turn, that is one after another, led to a drop in strain hardening and was equivalent to the effect of the monoslip. The metal matrix adopted to the last variant of texture development as a means for requiring minimal energy expenditure; this variant of deformation development explained details concerning strain hardening and resulting recrystallization in the considered samples.

The opinion is widespread in literature that by recrystallization of α-Zr, the main mode of texture changes consists of the rotation of the lattice by 30° about the basal axis [13, 15]. Some studies, however, do not support this opinion [16-18]. In the current chapter, an attempt was made to explain the discrepancies in experimental data concerning this question. In Fig. 13 PF{11$\bar{2}$0}, rolled and recrystallized α-Zr single crystals are shown at the highest deformation degrees. Though the deformation degrees of rolled single crystals were high and the stable distribution of basal axes already had time to form, the usual distribution of prismatic axes, characterized by regular arrangement of alternating in 60° maxima along the meridians of stereographic projection, almost did not develop. Since basal maxima are situated close to the centre of pole figure, prismatic axes are situated at the periphery of stereographic projection and are seen only partially. Smaller deformation degrees PF{11$\bar{2}$0} for rolled single crystals N1 and N 2 proved to be chaotic, so that the joint rotation by 30° of prismatic axes about the basal axis as a result of recrystallization could not be fixed.

This rotation means that in the rolled matrix there are recrystallization nuclei situated at the boundary between two deformed grains with close orientations, have a middle orientation and swallow grains of the matrix due to growing into these neighboring grains at equal rates. This mechanism can work only when the rolled matrix has developed a deformation texture, so that the rotation of prismatic axes by 30° is not the only possible situation in some fragments of the sample; instead, a more complex process takes place in a coordinated manner within the entire matrix. The misorientation of some newly originated nuclei by 30° with the deformed matrix can promote its growth, but it does not create the recrystallization texture in the absence of the developed rolling texture.

As for the rolled single crystal N 3 (Fig. 13e, f), this case deserves separate consideration. Due to the position of its basal axis near TD (Fig. 5j), the largest part of this single crystal deforms at first by means of a prismatic slip and only by $\varepsilon \approx 7\%$ the twinning is noticeable, overthrowing a portion of basal axes at the central segment of diameter ND-RD (Fig. 5m). PF{11$\bar{2}$0} of this single crystal after rolling (Fig. 13e) contains two branches: the left corresponds to the fraction deformed by the prismatic slip, i.e., to the developed rolling texture; the right corresponds to the twinned fraction, which is characterized by low strain hardening and a small number of recrystallization nuclei. Maxima in the right branch shifted up and down in correspondence to the position of twinned basal maxima near ND. The important feature of the rolled single crystal, as well as the rolled fine-grained polycrystal (revealed by their X-ray texture study) was the micro-scale of inhomogeneities, consisting in the approximate identity of results, which had been obtained through measurements of neighbouring regions. This result indicated that any region of the studied sample contained a statistically significant number of micro-fragments or grains with various orientations. As applied to the rolled single crystal N 3, this means, that its micro-fragment corresponded to both branches in PF{11$\bar{2}$0} (Fig. 13e), i.e., those deformed by prismatic slip, as well as small twins that can be their nearest neighbours in the microstructure of the sample. Therefore, as a result of recrystallization of this single crystal (Fig. 13f), the right twinned branch in PF{11$\bar{2}$0}, characterized by low strain hardening, proved to have been swallowed by nuclei from the left branch with the developed rolling texture. However, micro-fragments deformed by the prismatic slip after recrystallization did not indicate a clear rotation by 30°, since among the nearest neighbours of these micro-fragments small twins were present and the deformed matrix had been broken into particles with alternating characteristics, and coordinated rotation of this matrix proved to be impossible. Nevertheless, a number of weak signs of such future rotation were seen within the left branch of the pole density distribution in Fig. 13f.

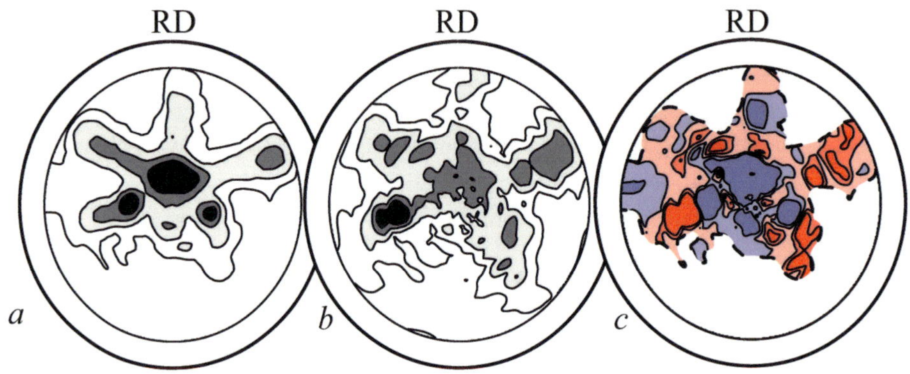

Figure 9. Recrystallization in cold rolled α–Zr single crystals N 2 (ε~30%): *a* – $PF(0001)_{roll}$, *b* - $PF(0001)_{recr}$; *c* – [$PF(0001)_{recr}$- $PF(0001)_{roll}$].

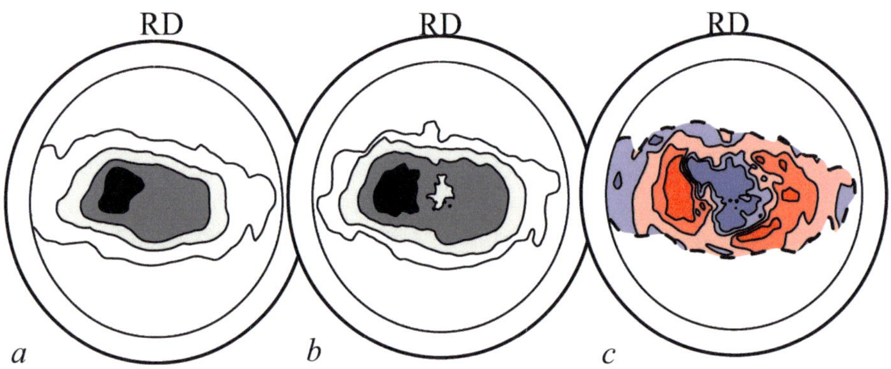

Figure 10. Recrystallization in cold rolled α–Zr single crystals N 2 (ε~70%) : *a* – $PF(0001)_{roll}$, *b* - $PF(0001)_{recr}$; *c* – [$PF(0001)_{recr}$- $PF(0001)_{roll}$].

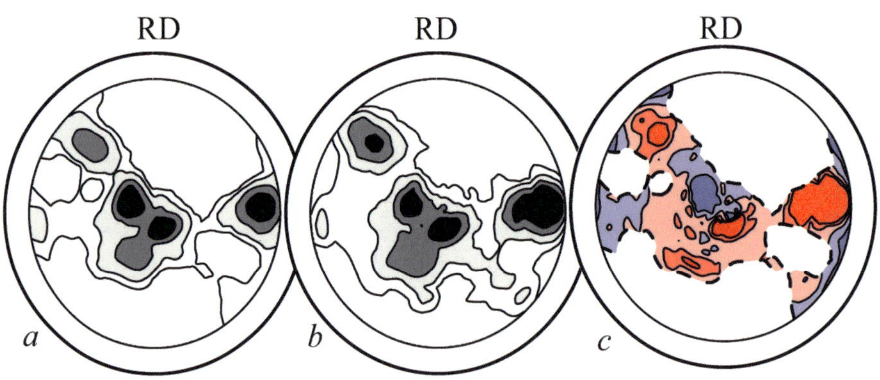

Figure 11. Recrystallization in cold rolled α–Zr single crystals N 3 (ε~30% (*a-c*) and ε~70% (*d-f*)): *a* – $PF(0001)_{roll}$, *b* – $PF(0001)_{recr}$; *c* – [$PF(0001)_{recr}$- $PF(0001)_{roll}$].

http://dx.doi.org/10.5772/60019

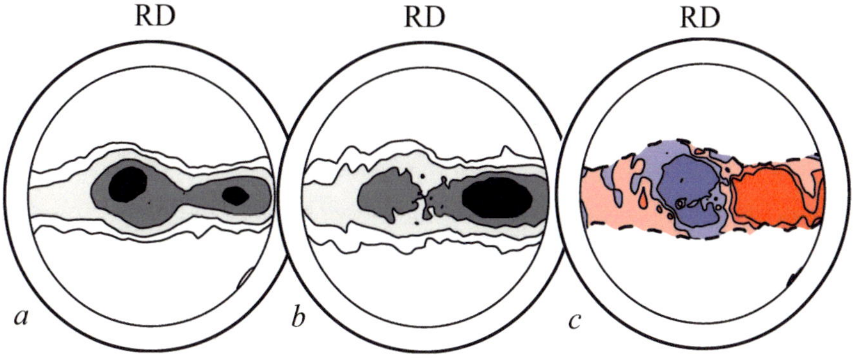

Figure 12. Recrystallization in cold rolled α-Zr single crystals N 3 (ε~30% (*a-c*) and ε~70% (*d-f*)): *a*– $PF(0001)_{roll}$, *b* – $PF(0001)_{recr}$; *c* – [$PF(0001)_{recr}$- $PF(0001)_{roll}$].

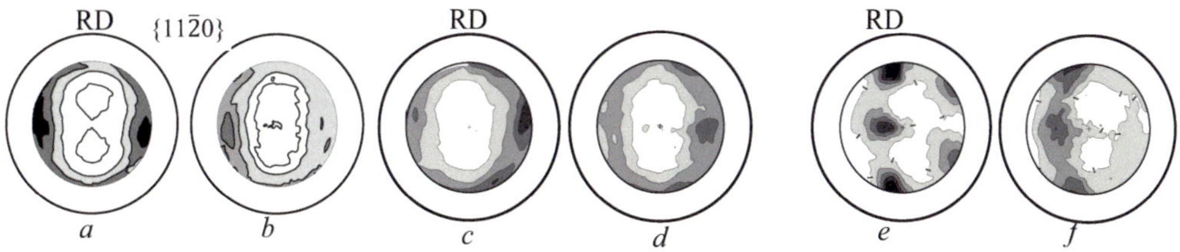

Figure 13. Texture changes through the recrystallization of rolled α-Zr single crystals N 1, ε=80% (*a, b*), N 2, ε=70% (*c, d*) and N 3, ε= 70% (*e, f*). (*a, c, e*): PF {11$\bar{2}$0} for rolled single crystals; (*b, d, f*):PF {11$\bar{2}$0} for rolled and annealed single crystals 580°C/3 h.

4.4. Recrystallization of polycrystalline α-Zr sheets in light of data, obtained for recrystallization of rolled α-Zr single crystals

The consideration of data regarding the recrystallization of a rolled α-Zr single crystal can be used for explaining the features of polycrystalline α-Zr sheets observed during their recrystallization. In Figs. 14 and 15, the results of X-ray diffractometric studies of iodide coarse-grained α-Zr rolled, respectively, by 50% and 70% and then recrystallized, are presented. In Figs. 16 and 17, analogous data are presented for α-Zr in a polycrystalline Zr-1%Nb alloy, rolled by 70% along and across the initial RD, respectively, and finally, recrystallized. In all Figures PF(0001), rolled samples are shown as (a), PF(0001) for recrystallized samples – as (b); GPF_{0002}– as (c); subtraction diagrams S(ψ, φ) - as (d) and diagrams of correlation PF(0001) - GPF β_{0002}- as (e). In Fig. 18, polycrystalline materials PF{11$\bar{2}$0} are shown for both rolled and recrystallized conditions.

According to models of texture formation in α-Zr [13, 19], the pole density in the central region of PF(0001) is conditioned by the input of grains experiencing twinning by deformation.

Subtraction diagrams show that under recrystallization in the corresponding regions of PF for the deformed samples, an increase of pole density was absent. The reasons for this effect in polycrystalline α-Zr sheets are the same as for rolled single crystals. The absence of growth of α-Zr grains with orientations corresponding to the central region of PF(0001) has been noted in previous studies dealing with the recrystallization of products from Zr-based alloys [8, 20]. However, only experiments with the rolling and recrystallization of α-Zr single crystals have yielded a direct demonstration of the effect as influenced by twinning on the strain hardening and the subsequent recrystallization of α-Zr crystallites.

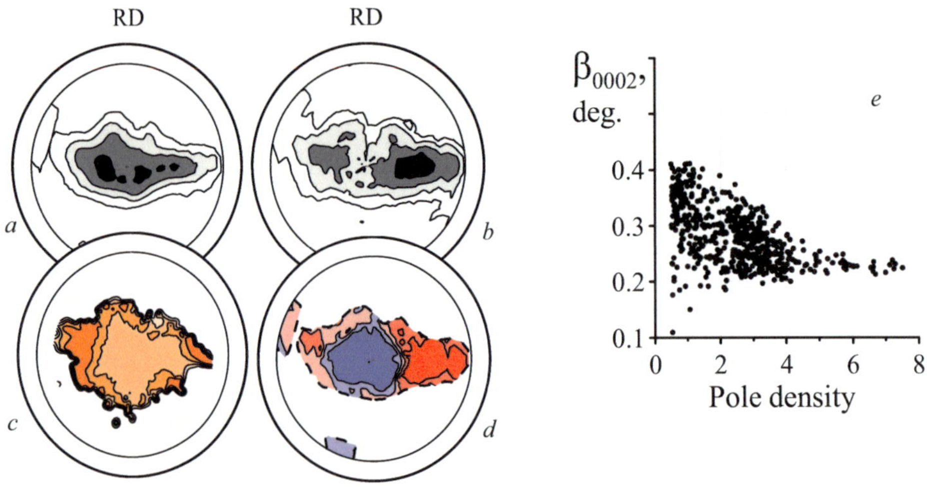

Figure 14. Recrystallization of polycrystalline rolled iodide α-Zr (ε~50%): *a* – $PF(0001)_{roll}$, *b* - $PF(0001)_{recr}$; *c* – GPF β_{0002}; *d* – $[PF(0001)_{recr}$- $PF(0001)_{roll}]$; *e* - diagram of correlation PF(0001) - GPF β_{0002}.

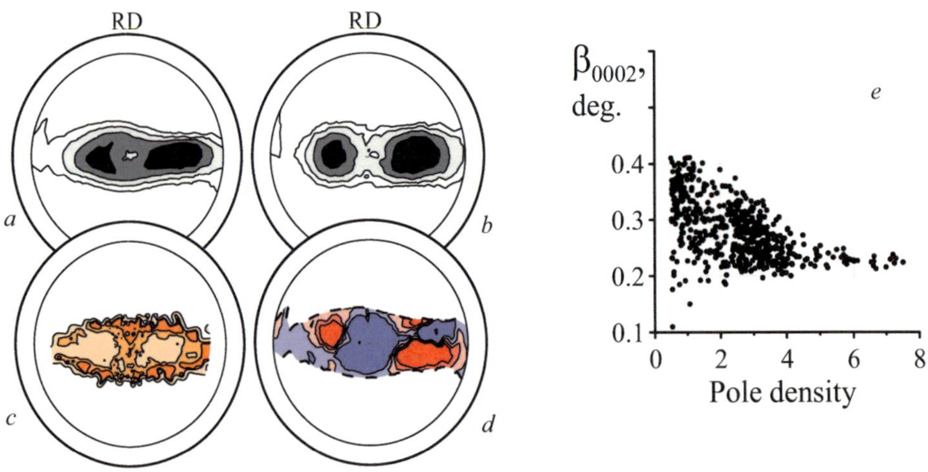

Figure 15. Recrystallization of polycrystalline rolled iodide α-Zr (ε~80%): *a* – $PF(0001)_{roll}$, *b* - $PF(0001)_{recr}$; *c* – GPF β_{0002}; *d* – $[PF(0001)_{recr}$- $PF(0001)_{roll}]$; *e* - diagram of correlation PF(0001) - GPF β_{0002}.

http://dx.doi.org/10.5772/60019

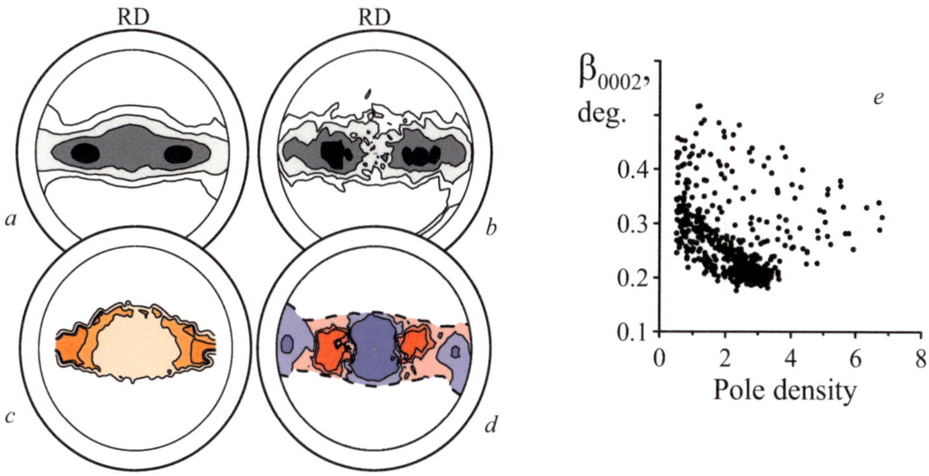

Figure 16. Recrystallization of Zr-1%Nb alloy rolled along initial RD (ε~50%): *a* – $PF(0001)_{roll}$, *b* - $PF(0001)_{recr}$; *c* – GPF β_{0002}; *d* – $[PF(0001)_{recr}$- $PF(0001)_{roll}]$; *e* - diagram of correlation PF(0001) - GPF β_{0002}

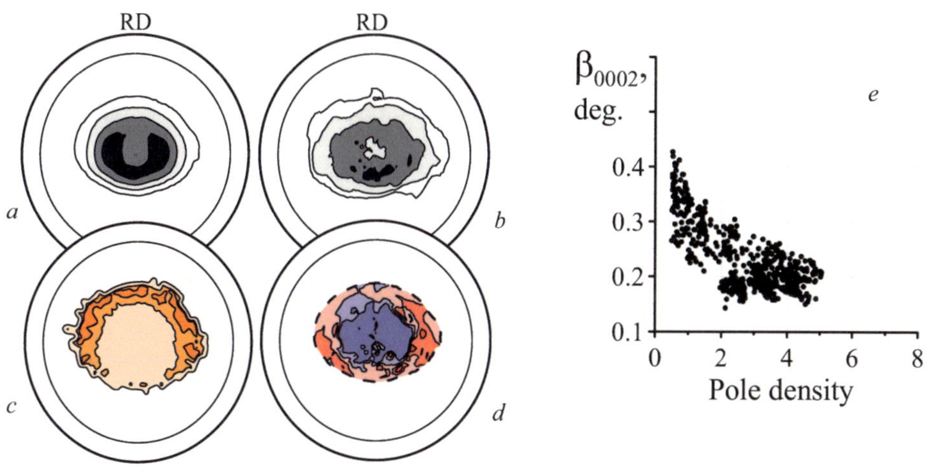

Figure 17. Recrystallization of Zr-1%Nb alloy rolled across initial RD (ε~50%): *a* – $PF(0001)_{roll}$, *b* - $PF(0001)_{recr}$; *c* – GPF β_{0002}; *d* – $[PF(0001)_{recr}$- $PF(0001)_{roll}]$; *e* - diagram of correlation PF(0001) - GPF β_{0002}

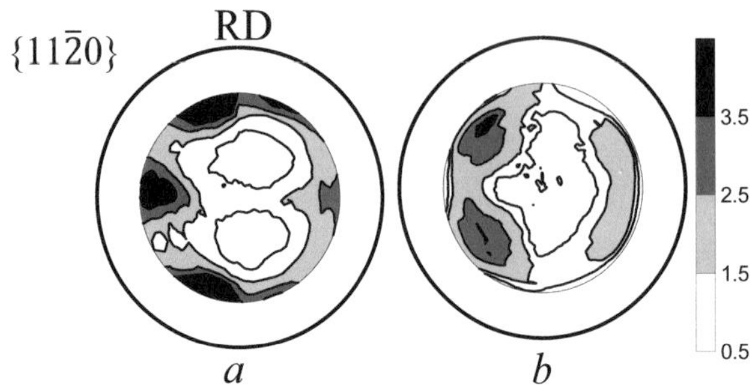

Figure 18. Texture changes through recrystallization of rolled Zr single crystals N 1 (a, b), N 2 (*c, d*) and N 3 (e, f). (a, c, e) DPF {11$\bar{2}$0} for rolled single crystals with deformation degrees: ε=80, 70 and 70%, respectively; (b, d, f) annealing 580°C – 3 h.

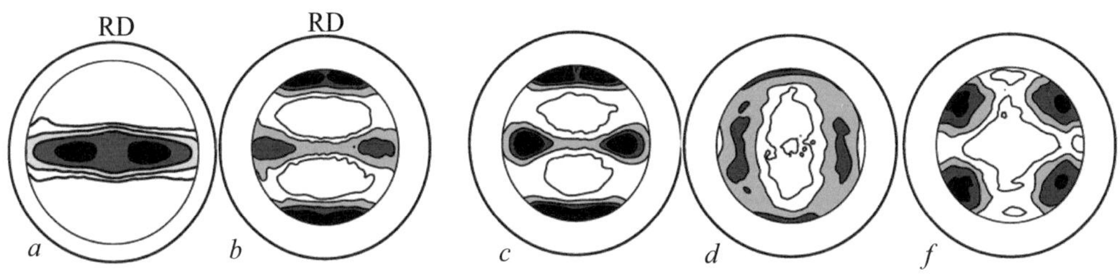

Figure 19. PF (0001) (a) and PF{11$\bar{2}$0} (b) for the rolled Zr–1% Nb plate; changes in PF{11$\bar{2}$0} due to recrystallization annealing upon heating at different rates: (c) 100, (d) 20, and (e) 2°/min.

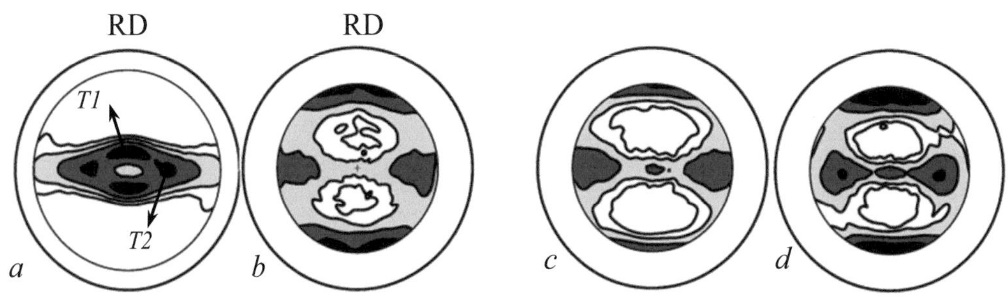

Figure 20. PFs (0001) (a) and {11$\bar{2}$0} (b) of the rolled Zr–1% Nb plate with texture components T1 and T2 in the initial state and changes in PF{11$\bar{2}$0} due to recrystallization annealing upon heating at different rates: (c) 100 and (d) 2°/min.

Shifts in the maxima rolling textures of polycrystalline Zr sheets by recrystallization are explained by the preferential growth of grains with basal axes orientations at the outward edges of these maxima. Here, there are more recrystallization nuclei and new grains that arise here absorb deformed grains localized within previous maxima textures (Fig. 14-17d). This shift occurs due to the inhomogeneity of strain hardening in the rolled samples, consisting first of all in its strengthening with increase of an angular distance of crystallite from texture maximum. This effect is clearly seen in Figs. 14-17 and has the same nature as bordering maxima textures with zones of increased strain hardening and intensive growth of recrystallized grains in rolled single crystals. The outward edge of different stable texture maximum slip systems, i.e., basal and prismatic systems, act together and since the volume of material to be deformed here is relatively small, simultaneous operation of both systems is possible. However, when moving inside the stable texture maximum, these slip systems begin to operate alternately in each point of the sample. As a result of recrystallization, new maxima grows at "slopes" of initial maxima in the rolling texture. Here, the strain hardening is sufficiently high for providing the preferential growth of arising nuclei, while at the same time the number of grains located at this segment of "slope" is sufficiently large for absorbing the entire deformed matrix. Obviously, this feature is typical for all metal material at the deformation stage and corresponds to the maxima arising in their deformation textures. Features of their recrystallization at the first approximation reflect the characteristics of strain hardening distribution, formed during the course of their rolling.

In terms of data related to the sample of Zr-1%Nb alloy rolled across initial RD, we observed that this sheet was characterized by the same features of strain hardening distribution and recrystallization development, the only difference being the space isotropy of constructed diagrams. As a result of symmetry, the obtained data were confirmed to be statistically safer.

In rolled Zr single crystal, deformed up to 80% by means of twinning and predominantly consisting of basal and pyramidal slips (Fig. 5b), after annealing at 580°C, a noticeable rotation of prismatic axes was not observed (Fig. 13). At the same time, in polycrystalline Zr rolled up to 80% (Fig. 15), after recrystallization, the sharp reorientation of prismatic axes became apparent (Fig. 18) by means of their rotation at 30° about the basal axis, coinciding with the correct texture maximum (Fig. 15b). It was evident that the rotation of prismatic axes about basal axes took place only in cases when recrystallized areas of α-Zr were deformed by rolling and with the inclusion of a prismatic slip, which according to Hobson's scheme [2], develops within wide regions of stereographic projection adjacent to TD. The development of stable α-Zr rolling texture is always realized by active participation of a prismatic slip, since the existence of texture maxima at a horizontal diameter of PF(0001) is conditioned by the mutually balanced operation of a slip in the prismatic and basal planes [6]. Only in a case where texture maxima are distanced from ND by a comparatively small angular distance, as in PF(0001) of the single crystal rolled to ~50%, can one certify that a prismatic slip did not participate in the formation of texture.

Increasing the heating rate up to attainment of the recrystallization temperature (see Fig. 19c, d and f) or the presence of additional components in the texture of the rolled sheet, for example, T1+T2 (Fig. 20), prevents the rotation of prismatic axes.

Analysis of data regarding the rotation of prismatic axes under annealing of rolled α-Zr allows for ascertaining the mechanism of its recrystallization, established earlier by considering strain hardening distributions using GPF β_{0002}. The presented data testify that in α-Zr by recrystallization, the mechanism is realized by combining two different aspects:

1. the rotation of basal axes in accordance with the strain hardening distribution, revealed with the help of GPF β_{0002} and predetermining the predominant growth of regions responding by their orientation to "slopes" of maxima in the rolling texture;

2. the nuclei of recrystallization formed in these regions during the course of their growth can cause (or not) 30º rotation of the matrix, depending on the participation of the prismatic slip in their deformation.

From nuclei arising in zones of increased strain hardening and having all possible orientations, the matrix chooses nuclei characterized by a 30°angle of misorientation at the boundaries as the best prospects for future growth. The increased heating rate by annealing of material and the presence in the deformed matrix of layers of grains with other orientations [8, 9] prevent such selection (Figs. 13, 20).

Dependence of the observed reorientation of prismatic axes on the heating rate of the sample can be explained in the following manner: the growth of recrystallization nuclei as part of the diffusion process requires a definite time period for the revealing of these nuclei, which are capable of swallowing the deformed matrix and to impose its own orientation as a result of the optimal angle of misorientation at grain boundaries.The presence in the deformed matrix of layers of grains with orientations of RD different from $\{10\bar{1}0\}$ also prevents rotation of the matrix by 30°, as does the violation of the optimal 30° misorientation at all boundaries of growing nuclei and grains of matrix [15].

5. Conclusions

1. Experiments with rolling and annealing of α-Zr single crystals helped to understand the mechanisms of α-Zr recrystallization in polycrystalline products, the absence of grain growth in regions of predominant twinning, rotation of the deformed α-Zr matrix about the basal axis and the conditions for the preferential growth of new grains.

2. Regions of the α-Zr matrix deformed by predominant participation of twinning were characterized by minimal strain hardening, by an insignificant number of recrystallization nuclei and a decreased trend towards grain growth, so that under recrystallization, these regions were swallowed by those that had higher strain hardening.

3. The necessary condition for 30º rotation of the deformed matrix about the basal axis through recrystallization of rolled α-Zr consisted predominantly of the participation of a prismatic slip in its plastic deformation.

4. In the course of α-Zr recrystallization the following mechanism for the preferential growth of definite regions was realized: these regions were characterized by relatively high strain hardening and orientations "slopes" of maxima in the rolling texture, where number of grains was sufficiently great in order to swallow the whole α-Zr matrix. Deformation of the α-Zr matrix at these "slopes" was realized by the simultaneous joint operation of basal, prismatic and pyramidal slip systems, whereas in central regions of maxima these systems operated alternately and provided minimal strain hardening.

http://dx.doi.org/10.5772/60019

5. By the quick heating of rolled samples with the purpose of their recrystallization, rotation of the deformed matrix by 30° about its basal axis did not take place, as in the case, when additional components were present in the rolling texture.

Author details

Yu Perlovich* and M. Isaenkova

*Address all correspondence to: yuperl@mail.ru

Department of Materials Science, National Research Nuclear University MEPHI (Moscow Engineering Physics Institute), Kashirskoe shosse, Moscow, Russia

References

[1] Perlovich Yu., Bunge H.J., Isaenkova M. Structure inhomogeneity of rolled textured niobium. Zeitschrifft fur Metallkunde, Materials Research and Advanced Techniques 2000; 91(2) 149-159.

[2] Hobson D.O. Textures in Deformed Zirconium Single Crystals. Transactions of Metallurgical Society AIME 1968;242 1105-1110.

[3] Perlovich Yu.A., Isaenkova M.G. Reorientation of α-Zr Crystallites under Rolling. Russian Metallurgy (Metalli) 1987;3 152-155.

[4] Isaenkova M.G., Perlovich Yu.A. Role of Twinning in Development of Deformation Texture in α-Zr. Physics of Metals and Metallography 1991;5 87-92.

[5] Matcegorin I.V., Rusakov A.A., Evstyukhin A.I. Analysis of Mechanism of Texture Formation in α-Zr by Use of IBM Modelling. In book: Metallurgy and Materials Science of Pure metals. Moscow: Atomizdat 1980;14 39-52.

[6] Isaenkova M. G. and Perlovich Yu. A. Kinetics and Mechanisms of Texture Formation in α-Zirconium upon Rolling. Physics of Metals and Metallography 1987; 107-112.

[7] Tenckhoff E. The Development of the Deformation Texture in Zirconium during Rolling in Sequential Passes. Met. Transactions A 1978;9 1401-1412.

[8] Perlovich Yu., Isaenkova M. Regularities of Recrystallization in Sheets and Tubes of Zr-Alloys. // Microstructural and Crystallographic Aspects of Recrystallization, Roskilde, Denmark, 1995, p. 371-376.

[9] Perlovich Yu., Isaenkova M. Recrystallization. Ed. K. Sztwiertnia, InTech, Croatia. p. 1-20.

[10] Perlovich Yu. Development of Strain Hardening Inhomogeneity during Texture Formation under Rolling of bcc Metals. In: Numerical Prediction of Deformation Processes and the Behaviour of Real Materials. Proceedings of the 15thRiso International Symposium on Materials Science. Ed. by S.Y. Andersen et al., Riso National laboratory, Roskilde, Denmark, 1994, p. 445-450.

[11] Perlovich Y., Isaenkova M., Bunge H.J. General Newly-Discovered Regularities of Structure Inhomogeneity in Textured Metal Materials. Materials Science Forum 2001;378-381 174-179.

[12] Isaenkova M.G., Perlovich Yu.A., Fesenko V.A. Modern Methods to Construct Experimentally Texture Complete Direct Polefigures by X-ray Data. Industrial Laboratory. Materials Diagnostics 2013;79(7) 1 25-32.

[13] Kocks U.F., Tome C.N., Wenk H.R. Texture and Anisotropy. Preferred Orientation in Polycrystals and their Effect on Materials Properties. Cambridge University Press, 1998. 676 p.

[14] Phillippe M., Phillippe M.J. Texture Formation in Hexagonal Materials. In: Texture of Materials ICOTOM-10, H.J. Bunge, ed. (Switzerland, Trans. Tech. Publ.), 1994, p. 1337-1350.

[15] Recrystallization of Metallic Materials. Ed. By F. Haessner, Dr.Riederer Verlag GmbH, Stuttgart, 1978, s. 293.

[16] Dewobroto N., Bozzolo N., Batberis P., Wagner F. On the Mechanisms Governing the Texture and Microstructure Evolution during Static Recrystallization and Grain Growth of Low Alloyed Zirconium Sheets. International Journal of Material Research 2007;97 826-833.

[17] Wagner F., Bozzolo N., Van Landuyt O., Grosdidier T. Evolution of Recrystallization Texture and Microstructure in Low Alloyed Titanium Sheets. Acta Materialia 2002;50 1245-1259.

[18] Gerspach F., Bozzolo N., Wagner F. On the Stability of Recrystallization Texture in Low Alloyed Titanium Sheets. Application of Texture Analysis, Ceramic Transactions 2008;201 593-600.

[19] Tenckhoff E. Deformation Mechanisms, Texture and Anisotropy in Zirconium and Zircaloy. ASTM, Special Technical Publication (STP 966), Philadelphia, 1988. 77 p.

[20] M.G. Isaenkova, Yu.A. Perlovich, V.A. Fesenko, O.A. Krymskaya, N.A. Krapivka, Soe San Tkhu. Regularities of Recrystallization of Rolled Single Crystals and Polycrystals of Zirconium and Alloy Zr-1%Nb. The Physics of Metals and Metallography August 2014;115(8) 756-764.

Chapter 4

Internal Crystallisation Method to Produce Oxide Fibres and Heat Resistant Composites

Sergei T. Mileiko

Additional information is available at the end of the chapter

http://dx.doi.org/10.5772/61024

1. Introduction

Most efforts in developing advance materials to be used in machines for transportation and power generation have been always directed towards enhancing the energy efficiency of the machines, which means decreasing fuel consumption per a unit of useful work. Researchers all over the world have done a lot to reach the goal including developing structural materials with higher specific characteristics to enhance the ratio of a payload to weight of the structure, improving aerodynamics, applying appropriate coatings on hot details, cooling them, etc. However, enhancing the use temperature of structural materials for hot parts of machines has been and will be always a most efficient way to improve the energy efficiency of transportation and power generation.

Most heat resistant materials for heavily loaded structural elements, those being Ni-based superalloys are approaching their physical limit set by their microstructural stability and melting points. Perhaps, this limit will be around 1100°C and a dependence of the maximum use temperature on year of the alloy development is asymptotically going to this limit.

The importance of developing heat resistant material beyond nickel superalloys can be clearly demonstrated by comparing the history of enhancing the use temperature of superalloys and the history of developing aviation jet engines expressed in fuel efficiency of them. Both histories are presented in Figure 1.

At present, at least three families of heat resistant materials of an enhanced use temperature are developing:

1. Ceramic matrix composites (CMCs);
2. Metal matrix composites (MMCs);

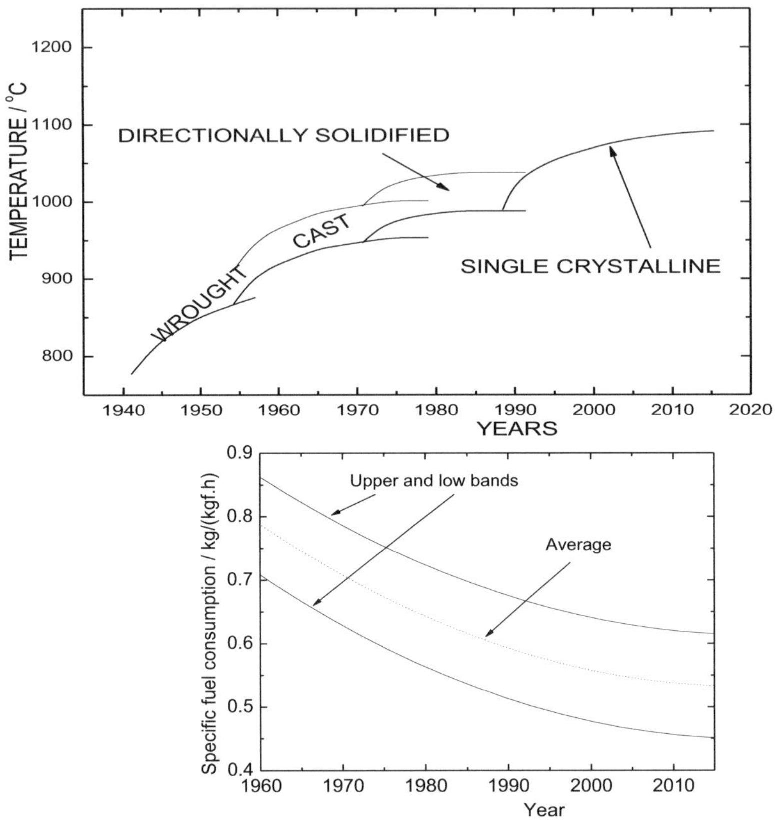

Figure 1. The growth of the use temperature of nickel superalloys [1] and decreasing the specific fuel consumption of aviation engines [2] with the time of their developing. The use temperature is determined by rapture stress 150 MPa at 1000 h.

3. Refractory metal alloys.

Active development of CMCs started with works by R. Naslain [3] and has now reached a stage of the active implementation [4]. The main efforts in this direction have been focused on silicon-carbide-fibres/silicon-carbide-matrix composites certainly due to a higher creep resistance of SiC fibre as compared to polycrystalline oxide fibres. Still, oxide/oxide composites have been developing to occupy an appropriate niche [5, 6, 7].

Metal matrix composites occur now to be in shadow of CMCs; however, it will be shown in this chapter that they have quite prospective future provided a number of the problems are solved.

The mainstream in the development of refractory metal alloys is alloying the matrix by silicon and some refractory metals to form complex silicides particles in the matrix. The main problem in the development of such alloys is to tailor their microstructure to reach an appropriate balance between creep resistance, fracture toughness and resistance to gas corrosion of the material to satisfy strict requirements set by the loading and heating conditions of a gas turbine blade surrounded by combustion products containing oxygen. Modern niobium alloys [8, 9], unlike those developed earlier [10] are characterized by high fracture toughness and oxidation resistance, which certainly allows using them at temperatures up to about 1200°C.

A further enhancement of the use temperature of heat resistant materials will be, perhaps, possible by appropriate tailoring molybdenum-based alloys. The present mainstream in this direction is developing alloys in the Mo-Si-B system [11]. Despite an essential enhancement of gas corrosion resistance of such alloys as compared to well-known molybdenum alloys [12], the balance between creep resistance and fracture toughness does not seem to have been reached yet [13].

A fundamental difference between metal alloys and metal matrix composites is a correlation between strength and fracture toughness of these two types of structural materials. Strengthening metal alloys, as a rule, limits dislocation movements and, hence, decreases crack resistance; reinforcing metal matrix with high strength **brittle** fibres yields obviously an increase in the strength and can yield an increase in the fracture toughness at the same time [1]. Therefore, development of MMCs promises better results than that of metal alloys.

To go on with the development of heat resistant MMCs, now some scientific and technical problems should be solved. Fabrication technology of appropriate fibres and composites with sufficiently high oxidation resistance are among the most important problems. The present chapter is to focus mainly on these items. We start with oxide fibres of single crystalline and eutectic structures and then proceed with discussing MMCs containing such fibres. Finally prospects of the internal crystallysation method to obtain new types of fibres and composites will be briefly discussed.

2. Fundamentals of the internal crystallisation method (ICM)

The only way to produce either single crystalline oxides or those with typical eutectic microstructure is to crystallize oxide melts. Three main methods to crystallise such fibres are well known:

1. Edge Feeding Growth (EFG);
2. μ-pulling down (μ-PD);
3. Laser heated pedestal growth (LHPG).

EFG was used to produce sapphire fibres for the first time by LaBelle and Mlavsky [14]. The method can actually be positioned within a concept of crystallising a melt by using a shaper formulated by Russian scientist Stepanov before the WWII [15]. Stepanov introduced a shaper to pre-determine a shape and size of the capillary column at the top of which the liquid/solid interface arises (Figure 2). Both a review of the corresponding techniques and discussion of the fibre growth parameters, structure and mechanical properties of sapphire fibres are presented in a paper dated by 1985 [16]. It appears that a stable growth takes place at rates of no higher than 0.5 - 1.0 mm/s, which makes productivity rate of the process low, so that the cost of the fibres is too high to use them in structural applications. The macrostructure and strength of sapphire fibres depend strongly on the crystallization rate.

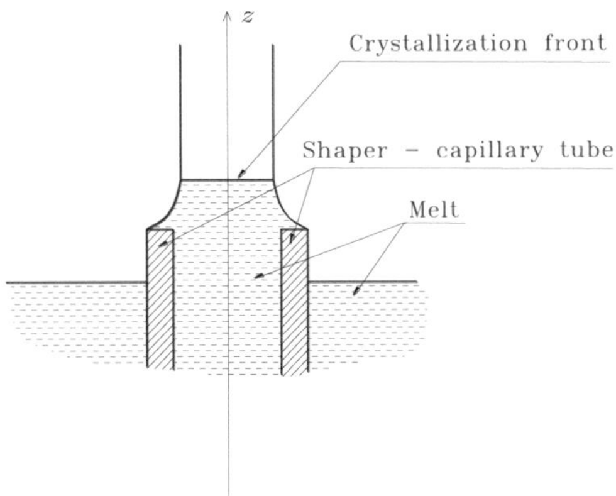

Figure 2. Capillary shaping-crystallization zone, a schematic view of Stepanov's concepts.

Developers of μ-PD method, Japanese researchers [17], did actually turn up down a classical scheme of EFG, which simplified slightly growth procedures. They claim that the usage of the μ-PD yields crystals with smaller thermal strains compared to other growth methods and using this method makes possible to grow crystals even from incongruent melts. Still, the method does not allow an essential decrease in the fibre cost as compared to EFG.

LHPG was used, perhaps for first time, to grow single-crystal Cr-doped Al_2O_3 fibres from a small source rods melted locally on its end by a CO_2 laser [18]. Then it has been used rather intensively to grow mainly optical fibres [19], although attempts to grow structural fibres are also known [20]. There are some advantages of the method: the absence of a crucible allows growing sufficiently pure crystals; a small volume of the melts yields an increase in both thermal efficiency despite of a low efficiency of the laser heating and and mass exchange around the process zone that would be useful in growing such materials as mullite, which is characterized by a complicated phase diagram.

The methods just described have been used to produce continuous oxide fibres of high quality, which are effectively used in various non-structural applications (laser optics, sensors, scintillations, etc.). At the very beginning of the composite era, in 70s of the last century, there was a hope to use sapphire fibres produced by EFG in structural applications so their high temperatures strength was a subject of an intensive study [21, 22, 23, 24]. However, high cost of the fibres produced by crystallisation of the melt has eventually led to losing interest to such fibres as reinforcement in heat-resistant composites technology.

Internal crystallisation method invented by Mileiko and Kazmin was described for the first time about 30 years ago [25, 26]. Since then a main scheme of the method as well as some variation of it has been published in a number of the papers (see references [27, 28, 29, 30, 31,

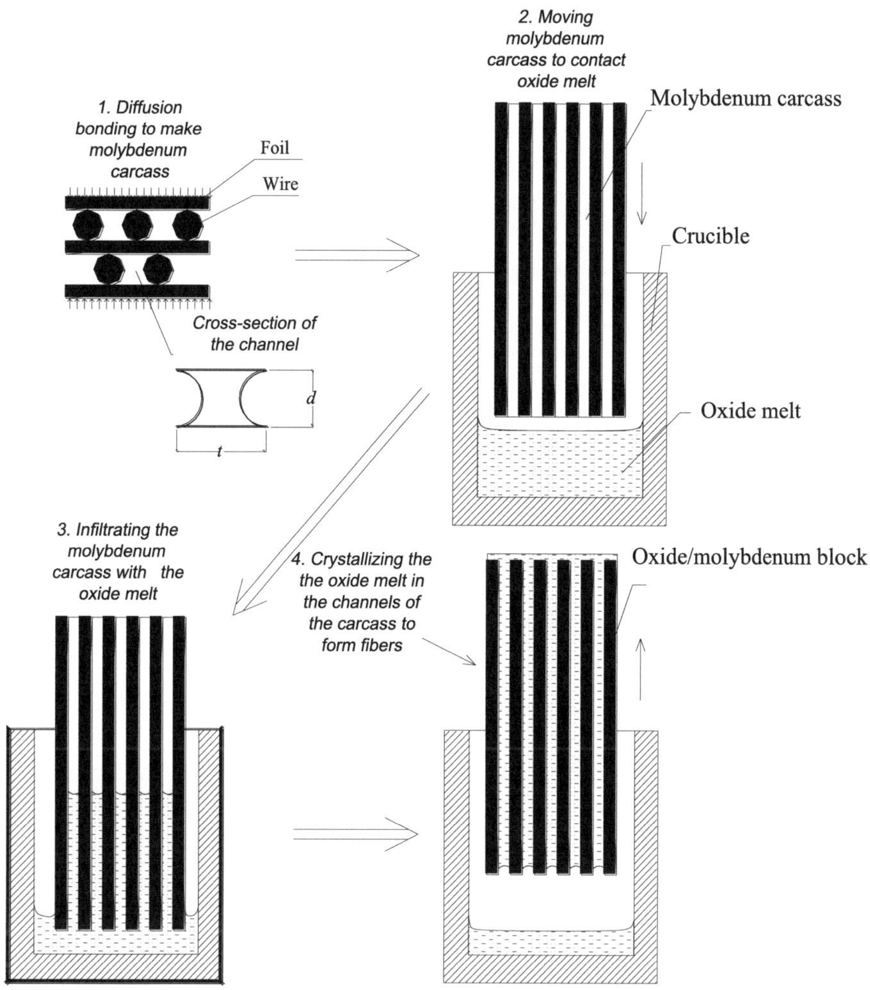

Figure 3. Schematic of the internal crystallization method (ICM).

32, 33]). All these papers are concerned with oxide fibres. However, recently it was shown a possibility to produce a larger family of the fibres. I particular, in Section 7.2 prospects of the internal crystallisation method composites with silicide fibres will be described.

A schematic of the method is presented in Figure 3. A molybdenum carcass with continuous channels in it, which is prepared by diffusion bonding of an assemblage of the wire and foil (step 1 in Figure 3), is infiltrated with an oxide melt (steps 2 and 3) by the capillary force. The melt is then crystallised in the channels to form fibres in an oxide/molybdenum block (step 4). This is a main scheme of the ICM, which can be varied to attain a particular goal. For example, to obtain sapphire fibres of a homogeneaous crystallographic orientation, a seed oriented in an appropiate manner is used as shown in Figure 4.

Figure 4. A set of the molybdenum carcass with sapphire seeds at the top of it.

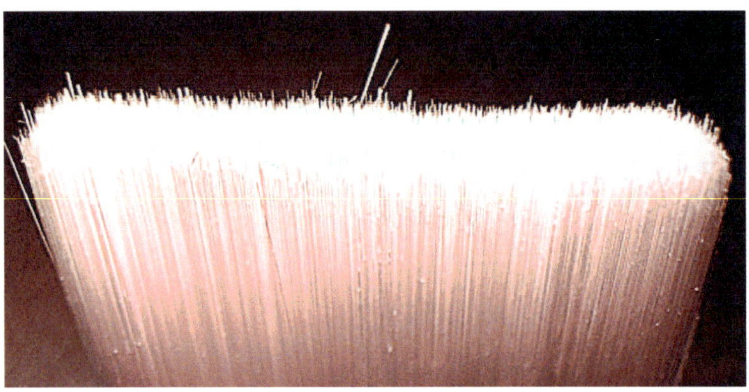

Figure 5. A bundle of sapphire fibres obtained by using ICM.

Finally, molybdenum is dissolved in a mixture of acids to get a bundle of the fibres (Figure 5). At the present time, the length of the fibre bundle can be up to 250 mm. Fibre volume fraction in a molybdenum block is normally about 0.4. The fibre has a cross-sectional shape close to that shown shematically in Figure 3, a characteristic size of it is usually 50 to 300 μm. A number of the blocks, which can be crystallised in one process as shown in Figure 4 is limited only by the size of the thermal zone of the crystallising machine. A set shown in Figure 4 contains after the crystallisation process neary 40,000 fibres with characteristic cross-sectional size of 100 μm.

http://dx.doi.org/10.5772/61024

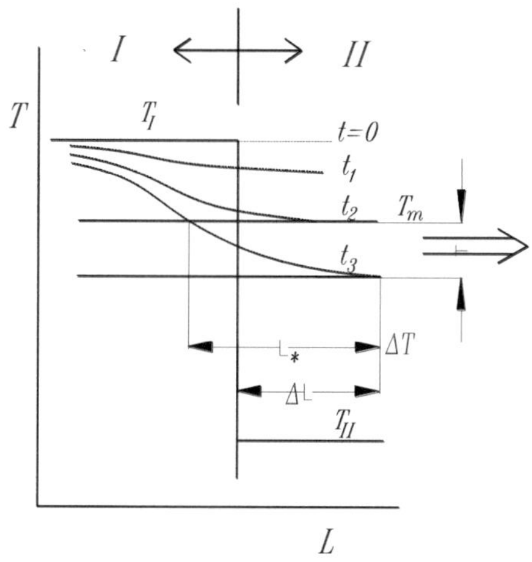

Figure 6. Schematic representation of temperature profiles in the crystallization zone.

3. Structure and mechanical properties of fibres produced by ICM

The microstructure and mechanical properties of fibres produced by ICM are determined mainly by the crystallization process of the fibre in the channels of the auxiliary molybdenum matrix.

3.1. Microstructure of ICM-fibres

A qualitative model of the process is suggested in references [26]. Suppose we have a furnace with two temperature zones "I" and "II" (Figure 6) and there is no direct thermal exchange between them. After the infiltration, the specimen is located within Zone I and its temperature is T_I, which is higher than melting point T_m of the fibre material. Now if the specimen is moved instantaneously in the direction of the arrow by distance ΔL, the temperature profile along the specimen starts changing as shown by lines marked with $t = 0$, t_1, t_2,... When some overcooling, say ΔT, is reached at time t_3, spontaneous crystallization within length L^* occurs. As a result of the process a solid-liquid boundary in each channel is emerged. Then the events can go on in a number of the ways depending on a concrete material.

3.1.1. Sapphire

It is well known that sapphire has no preferable crystallization direction. Hence, if a channel effective diameter is sufficiently small then each crystal on the solid side of the solid-liquid boundary can serve as a seed for further crystallization and the fibres obtained have random crystallographic orientations [26]. It is important to note that the probability of having a single crystal occupying the whole channel transverse section increases with the cross-sectional sizes of the channels decreases. A layer of the fibres in a tensile specimen obtained by using ICM, which illustrates the scheme described is shown in Figure 7.

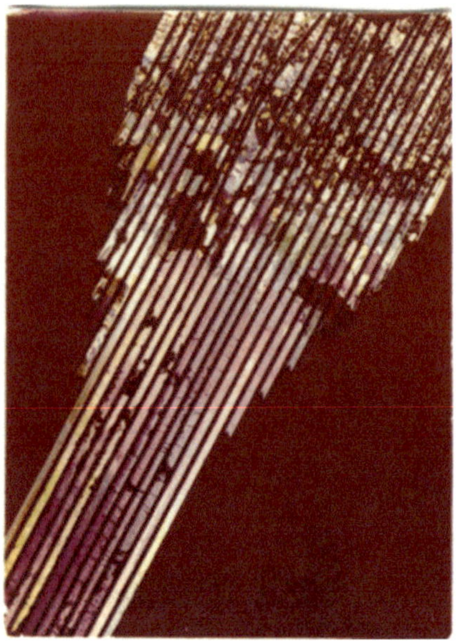

Figure 7. A layer of the sapphire fibres at the top of a sapphire/molybdenum tensile specimen obtained by ICM.

3.1.2. Yttrium aluminium garnet (YAG)

Fabrication technology of bulk crystal of YAG ($Y_3Al_5O_{12}$) normally includes high temperature annealing of crystals to restore YAG since cooling an overheated melt of the $Y_3Al_5O_{12}$ composition yields decomposition of YAG into alumina and perovskite $YAlO_3$. This stage of the process while obtaining single crystal YAG fibre by using ICM should be avoided because it can cause additional stresses in fibres as recombination of YAG is accompanied by a volume change. Hence, while crystallising garnets by using ICM an overheating of the melts is to be excluded [31, 34].

3.1.3. Mullite

Mullite has been known as most creep resistant oxide [35, 36]. That is why, obtaining single crystal mullite fibres looks as a challenge for the composite community. However, some reasons including a complicated phase diagram of the alumina-silica system [37] set serious problems in crystallising the fibres from the melt while using either EFG or LHPG crystallisation methods.

Using ICM allows crystallising single crystal mullite fibres [34,35, 38]. Crystallisation of a raw melt composed of a mixture of alumina and silica in the channels of a molybdenum block is a more complicated process than that of sapphire. The process was studied in details in references [34,35]. The fibre microstructure is determined by the crystallization process.

To study specific features of crystallisation of the melt containing a mixture of Al_2O_3 and SiO_2, a number of the raw compositions, $n\ Al_2O_3{:}SiO_2$, were used with the value of $n = 1.5$, 1.8 and

http://dx.doi.org/10.5772/61024

2.05. As in the case of sapphire, the fibre microstructure is changing in the upper portion of the block where crystallisation starts and then remains fairly unchanged. The steady state growth yields the fibres of nearly single-crystalline structure. Typical cross-sections of the fibres corresponding to lowest and highest values of n are presented in Figure 8.

Consider first a fibre obtained from the raw mixtures corresponding to $n = 2.05$ (Figure 8). The X-ray microanalysis shows that the black phase in SEM-pictures has atomic ratio Al:Si slightly higher than that in the precursor ($n = 2.05$). Despite of an excess of alumina in a fibre on the average, in the sharp corners of some fibres a "white" phase can be seen. With the value of n decreasing, an excess of silica in the raw material yields larger areas occupied by the "white" phase (Figure 8a). The "white" phase occurs to be enriched with silica and it is certainly a glassy phase. Again, a composition of the "black" phase corresponds to mullite. Impurities contents in the two phases of the fibres indicate that a growing mullite crystal pushes the impurities, which exist in the melt, to a silica enriched melt that solidifies at a lower temperature to form inclusions of the glassy-phase.

The microphotographs presented and the data of the The X-ray microanalysis [34] reveal some important features of the fibre microstructures:

1. The composition of mullite in the fibres is close to $2Al_2O_3$:SiO_2 independent of the initial composition of the precursor. A rather broad field of the existence of mullite in various phase diagrams published explains deviations from the exact 2:1 molar ratio.

2. Mullite is optically transparent, which is characteristic to single crystal. The content of impurities in the mullite phase does not reach the sensitivity of the experimental method.

3. Inclusions of the glassy phase are located mainly at the fibre periphery.

Optical microscopy [32] shows that the fibers consist of a mosaic of single crystal areas up to about 5 mm in length. The *c*-axis of the single crystal individuals are misalligned up to ±3° with respect to the fiber axis.

The events occurred in the volume of spontaneous crystallisation in the case of mullite fibres are similar in some details to those while crystallising sapphire fibres and different from it in other details. Observing changes of the microstructure of mullite fibres crystallised from the melts with $n < 2$ along the length of an oxide/molybdenum block provides an insight into the growth mechanisms of mullite that yield a microstructure of the fibre on the steady state growth portion. The microstructure of fibres at the top of an oxide/molybdenum block is presented in Figure 9. One can see a number of the mullite crystals divided by silica-based phase. For example, in an area marked by a circle, three crystals are seen. The direction of all crystals along the fibre axis coincides certainly with *c*-axis of the mullite crystal. However, their angular orientations can vary.

Crystallisation proceeds with growing some crystals of various angular orientations. The process is accompanied with pushing a liquid, which is to form the glassy phase, towards the oxide/molybdenum interface. Obviously, the crystallisation process depends on many parameters such as pulling rate, temperature gradients, melt composition, etc.

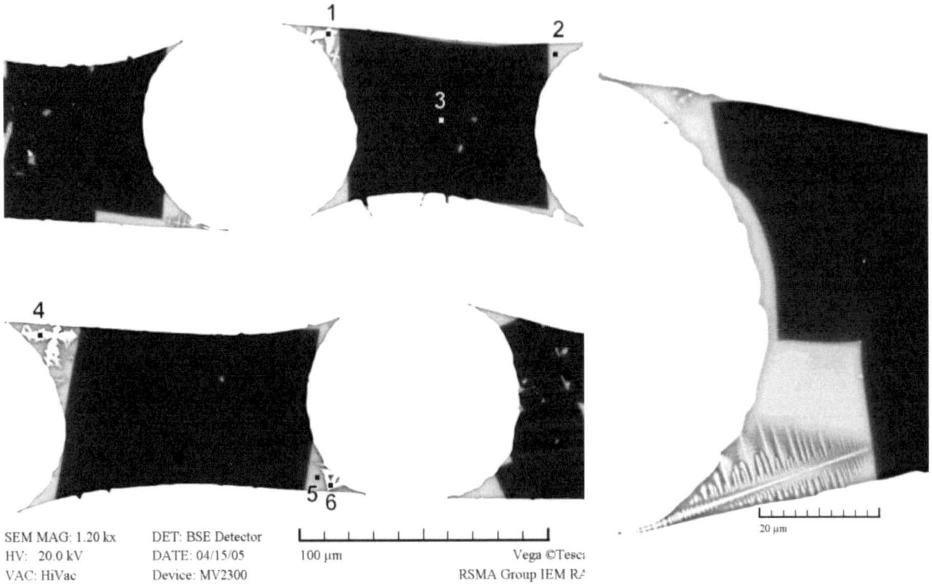

(a)

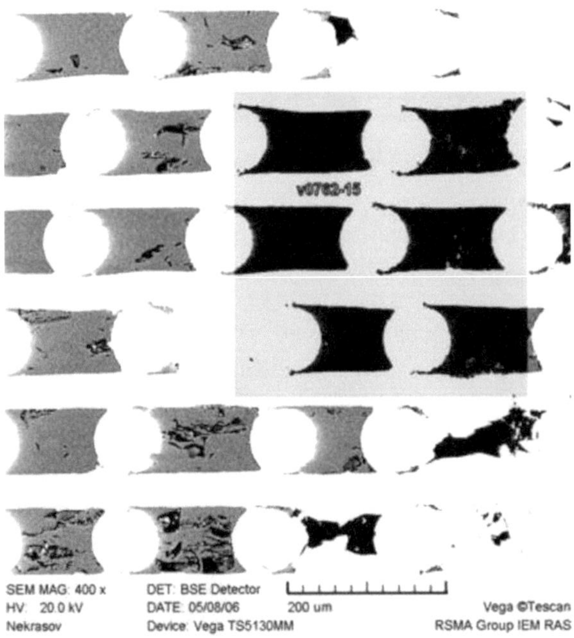

(b)

Figure 8. Typical cross-section of fibres in a region of the steady-state growth obtained from the melt of mixtures $1.5Al_2O_3:SiO_2$.(a) and $2.05Al_2O_3:SiO_2$ (b). Black areas are mullite; white ones are glassy-phase.

http://dx.doi.org/10.5772/61024

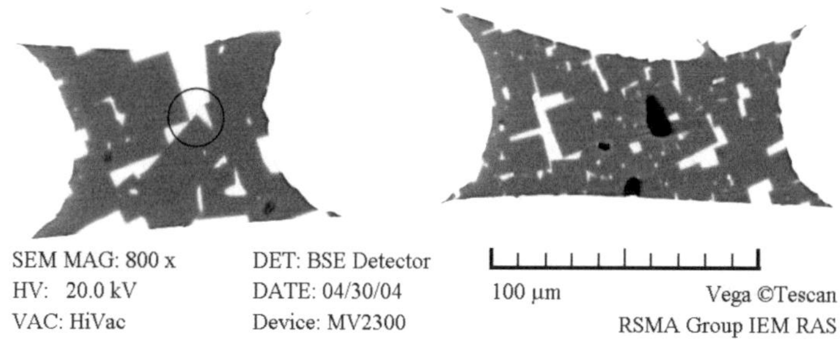

Figure 9. The microstructure of fibres in top of a fibre bundle shown in Figure 8a (raw mixture corresponds mixture $1.5Al_2O_3{:}SiO_2$).

3.1.4. Oxide eutectics

Special features of crystallization process in the channels of a molybdenum carcass determine particularities of the eutectic microstructure of the fibres. Perhaps, most important feature is a non-plane solid-liquid boundary in the channels due to heat sinks into molybdenum carcass. Still, characteristic microstructure of the eutectics is observed in the fibres obtained by using ICM, Figure 10, 11 and Figure 12 [39]. A non-homogeneity of the microstructures is a result of the features of crystallization process mentioned above. The dependence of the characteristic microstructure size of the Al_2O_3-$Y_3Al_5O_{12}$ eutectic fibres on the pulling rate related to crystallysation rate (Figure 11) is certainly close to a usual one.

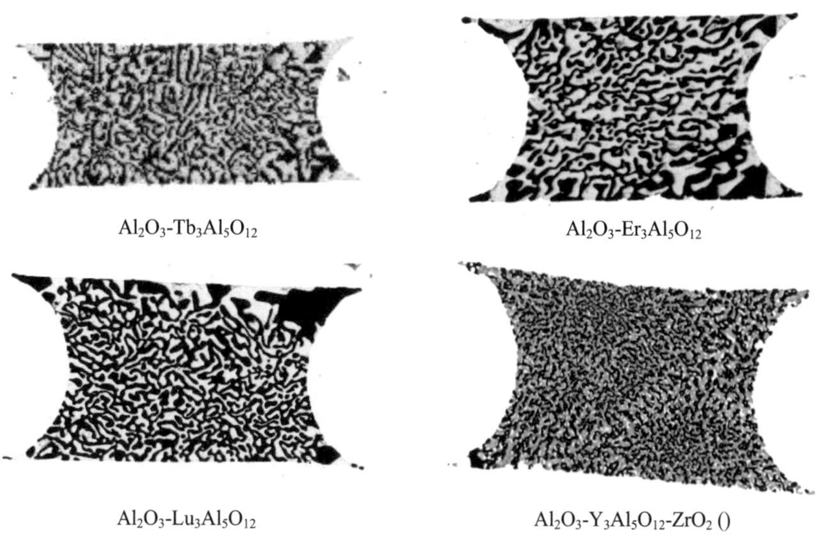

Figure 10. The microstructures of Al_2O_3-$Re_3Al_5O_{12}$ (Re = Tb, Er, Lu) and Al_2O_3-$Y_3Al_5O_{12}$-ZrO_2,eutectic fibres crystallised with pulling rate 1.7 mm/min. Characteristic size of the fibre cross-section (distance between the plane surfaces) is ~65 mm.

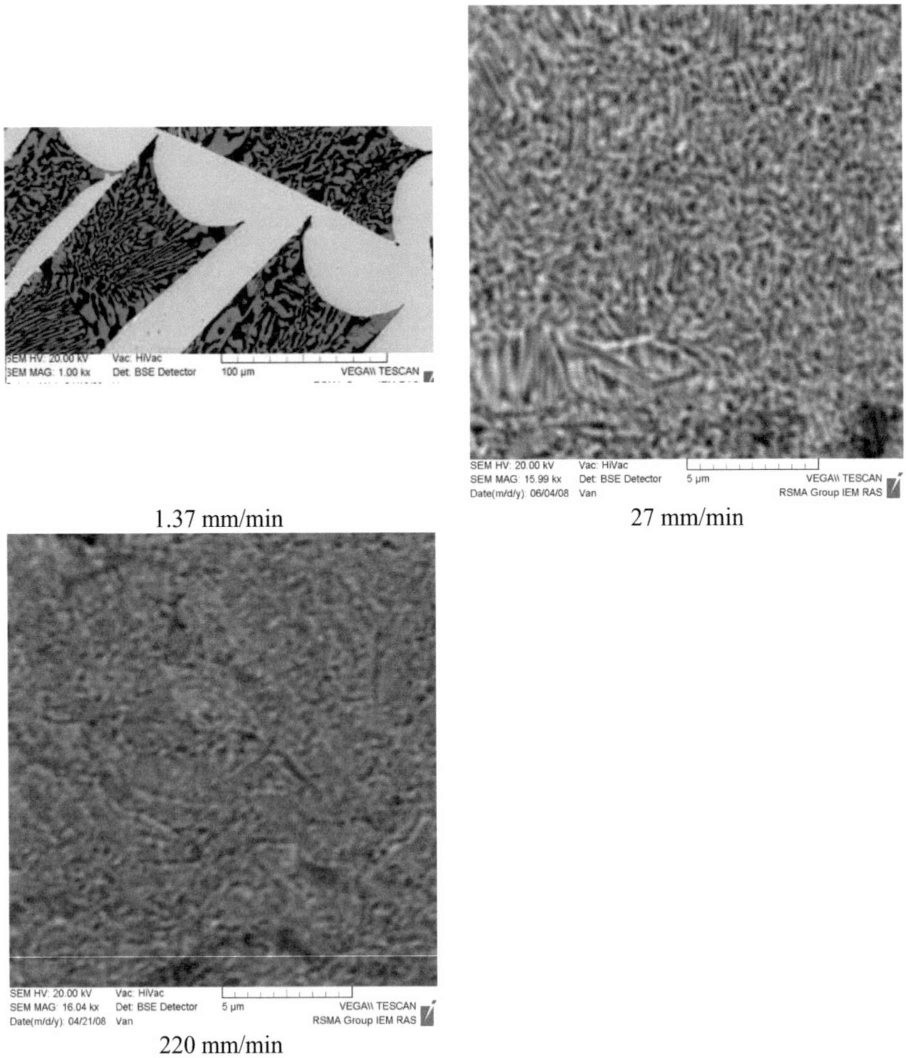

Figure 11. The microstructures of Al_2O_3-$Y_3Al_5O_{12}$ eutectic fibres crystallised with various pulling rate.

3.2. Room temperature strength of fibres produced by ICM

First of all, it should be noted that the strength of fibres produced by using ICM as well as that of all structural fibres depends on their surroundings: the results of testing separate fibres can drastically differ from the effective strength of the fibre in a matrix [40].

Another point to be mentioned concerns methods of measuring strength This is important since the fibres have an unusual shape of the cross-section.

The strength of separate fibres is measured by looping a fibre around a series of the rigid cylinders of successively smaller diameters and calculating the bending strength [41]. The

http://dx.doi.org/10.5772/61024

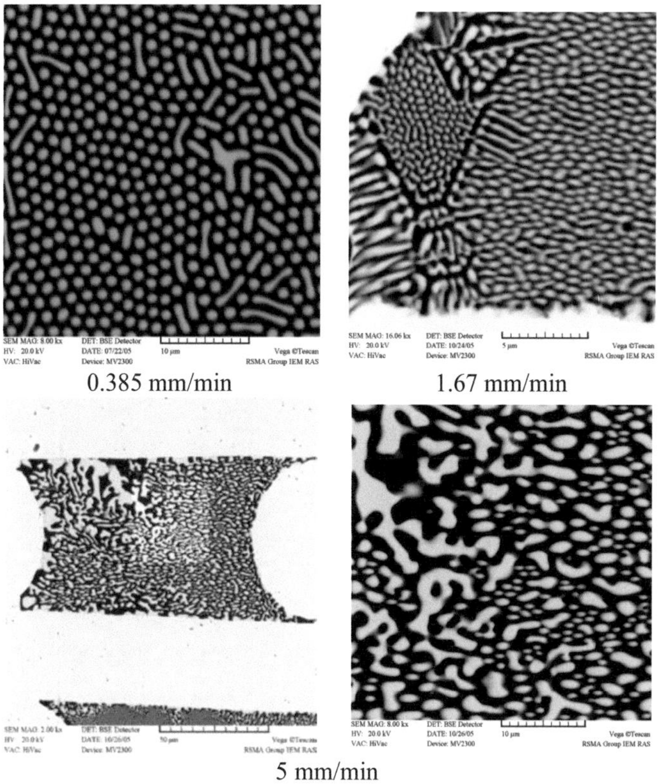

Figure 12. The microstructures of Al_2O_3-$GdAlO_3$ eutectic eutectic fibres crystallised with various pulling rates.

maximum fibre stress corresponding to the rigid cylinder of radius R is calculated according to

$$\sigma = E\frac{d}{2R} \tag{1}$$

where E is the Young's modulus of the fibre material and d is the distance between the plane surfaces of a fibre. To a first approximation, σ is the fibre bending strength at a length equal to the average distance between the fibre breaks.

Results of measuring sapphire fibre bending strength are presented in Figure 13. It should be pointed out that coating the fibre with a thin layer of silicon oxi-carbide by uing CVD process yields an essential increase in the fibre strength. This effect is a result of healing fibre surface defects by the coating material.

An effect of stress concentration in the vicinity of a defect in the fibre can be enhanced as a result of temperature gradients. Hence, annealing fibres in a furnace either without or with small temperature gradients leads to an increase in the fibre strength as shown in Figure 14 and Figure 15 for sapphire and single crystalline YAG fibre, respectively.

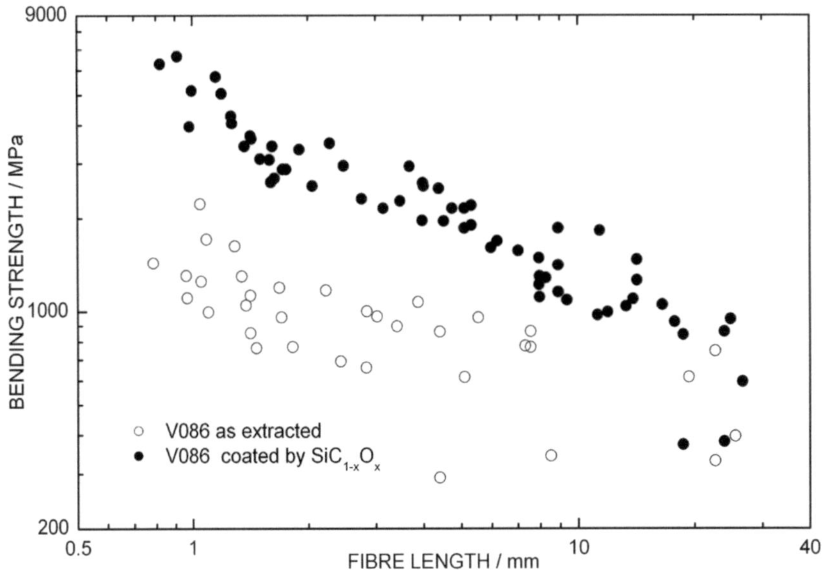

Figure 13. Bending strength of as-extracted and coated sapphire fibres versus fibre length. The fibres are coated by a layer of $SiC_{0.99}O_{0.01}$ of thickness 4-6 μm [33].

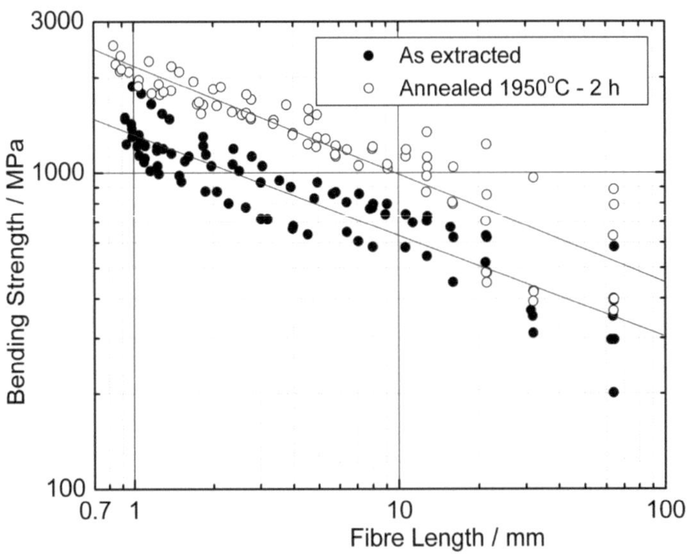

Figure 14. Scale dependence of sapphire fibre strength, as extracted and annealed to decrease permanent stresses. All the fibres are from one batch [33].

The microstructure and strength of eutectic fibre depends on the crystallization rate, with the increasing a characteristic size of the eutectic microstructure is decreasing as shown in Figure

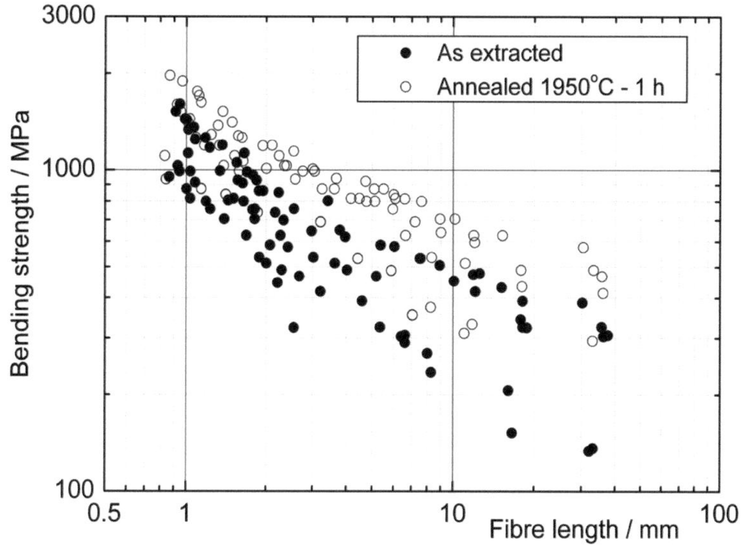

Figure 15. Bending strength versus fibre length for as extracted and annealed YAG fibres. All the fibres are from one batch. [36].

11 and 12. Hence, the strength of eutectics is expected to increase. However, the experiments yield a strength/crystallisation-rate dependence with a maximum at a some value of crystallization rate (Figures 16 and 17). This phenomenon remains to be studied.

3.3. High temperature strength of fibres produced by ICM

The measurements of high temperature strength of ICM-fibres [33] confirm early finding that single crystalline oxides retain their nearly room temperature strength up to high temperatures. Obviously, it is difficult to perform high temperature tests of ICM-fibres in the same manner as it is done at room temperature. However, the method of producing the fibres provides a way to test fibres that is to test oxide/molybdenum specimens. This way has been used from the very beginning of the developing of the internal crystallisation method.

Early experiments with oxide-fibre/molybdenum-matrix composite specimens described in [26,27] were conducted in tension. Then most tests have been performed under bending; such tests can be done with simple specimens. To calculate the fibre strength in molybdenum carcass with volume fraction of the channels of about 40% one needs to know mechanical properties of molybdenum used for the carcass in fully recrystallised state. The necessary data have been obtained in tensile tests, see Figure 18, and the strength/temperature dependence is approximated by a polynom of the 3rd order. The strength data obtained for some fibres are presented in Figure 19 and Figure 20. One can see that (i) the effective strength of ICM-sapphire fibre in the molybdenum matrix does not significantly differ from that for the fibres obtained by EFG method, (i) the fibre strength obtained in bending tests is slightly higher than that obtained in tensile tests, which should be expected.

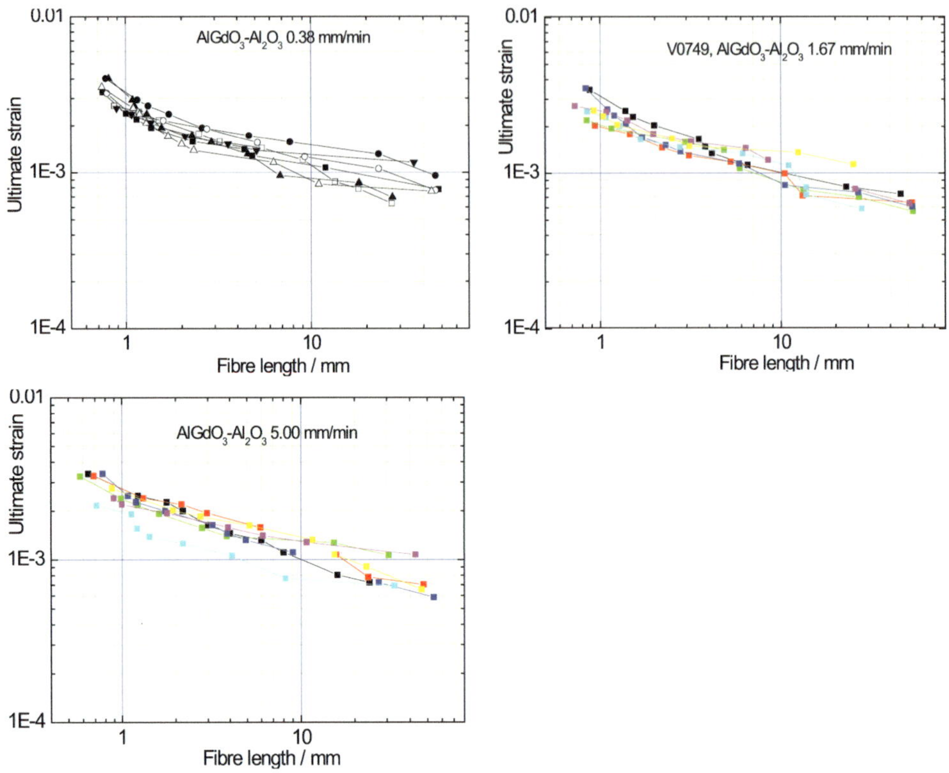

Figure 16. Ultimate strain versus fibre length for Al_2O_3- $GdAlO_3$ eutectic fibres crystallised with various pulling rates.

3.4. Creep characteristics of fibres produced by ICM

Creep characteristics of ICM-fibres are measured by testing oxide-fibre/Mo-matrix composites obtained at step 4 in Figure 3. It should be noted that oxide melts wet molybdenum excellently and the fibres have been crystallised in the channels of the molybdenum matrix, so the fibre/matrix interface is considered to be an ideal; hence, the fibre strength is not affected by possible surface flaws. Therefore, testing fibres in their mother matrix yields their virgin properties. Any further processing of the fibres changes their properties. Note that the fibre characteristics are easily obtainable as a contribution of the matrix, which is fully recrystallised molybdenum, to the creep resistance of oxide/molybdenum composites even at the lowest temperature, 1000°C, is negligible, less than 10 MPa [27].

An analysis of the bending tests performed under a varying load yields approximate creep characteristics of the composite in a particular specimen [42]. This allows excluding a number of the factors, which determine a usual scatter of the creep data of a batch of the material. Performing such tests yields a characterisation of each specimen by its own set of the creep characteristics. The analysis is based on the following assumptions:

1. A composite is characterised by identical creep behaviour under tension and compression

http://dx.doi.org/10.5772/61024

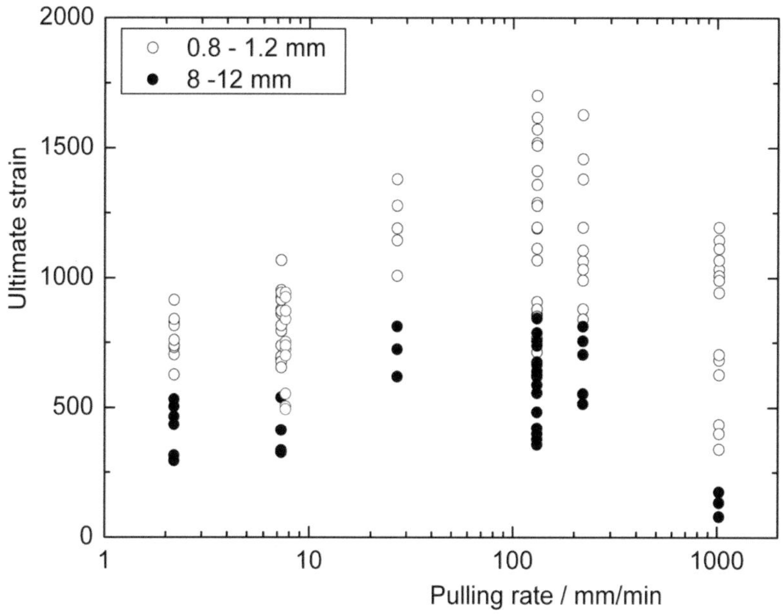

Figure 17. Dependence of the strength of Al_2O_3-$Y_3Al_5O_{12}$ eutectic fibres on two length bases on crystallization rate. The value of the Young's modulus of the eutectic is accepted to be 350 GPa

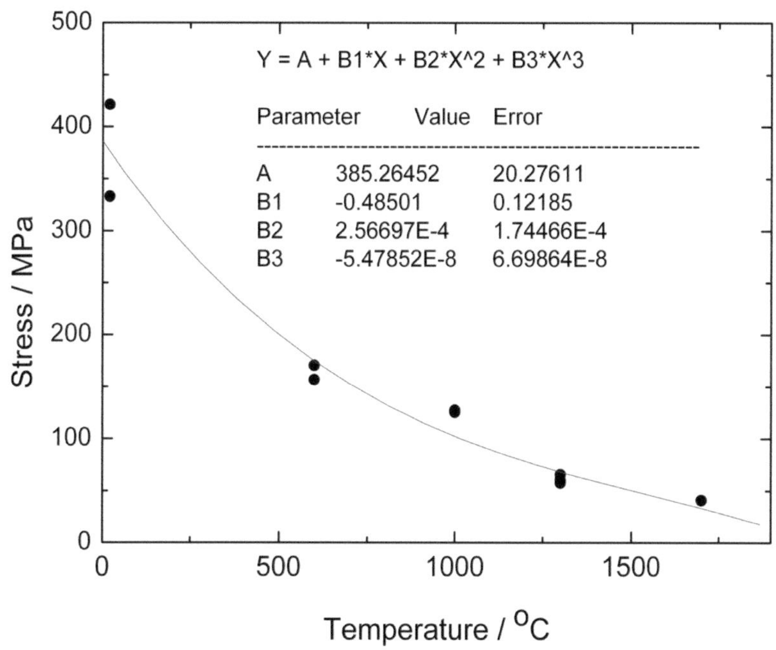

Figure 18. The temperature dependence of the tensile strength of a molybdenum carcass subjected to heat treatment at 2050°C for 30 min [26]. Approximation by a polynom of 3rd order with values of the constants shown.

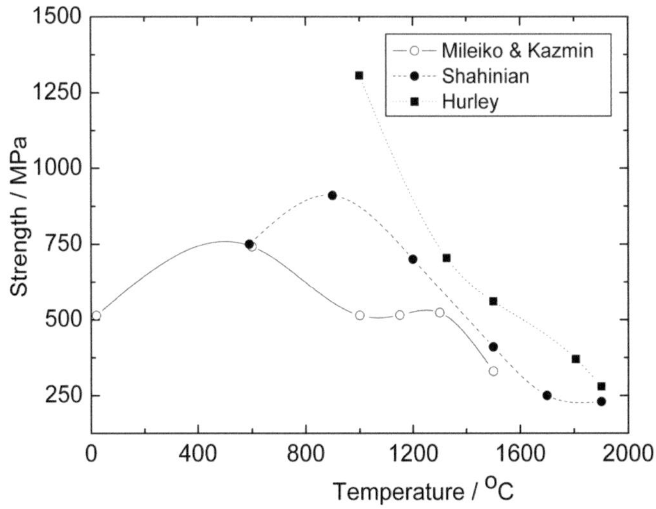

Figure 19. Temperature dependence of the tensile strength of sapphire fibres obtained by both the ICM and EFG-method. Experimental data are after Mileiko and Kazmin [27], Hurley [22] and Shahinian [21]. The present data are obtained by testing sapphire/molybdenum specimens in 3-point bending.

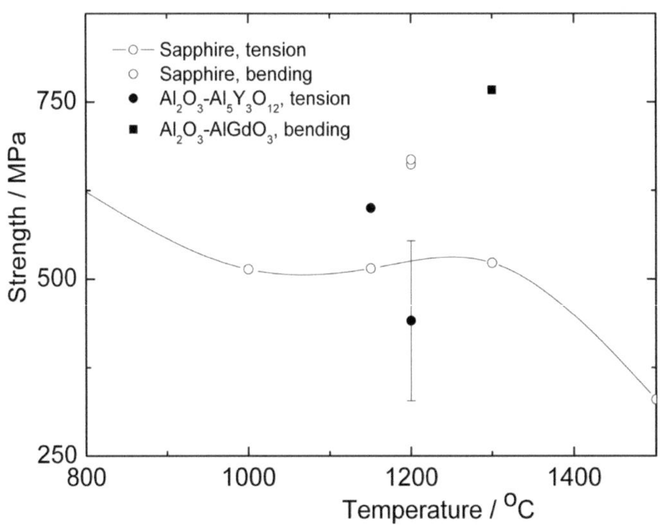

Figure 20. High temperature strength of some oxide fibres produced by ICM. The data were obtained by testing oxide/molybdenum specimens [33].

2. The creep law of the material is

$$\dot{\varepsilon} = \eta_n \left(\frac{\sigma}{\sigma_n} \right)^n \tag{2}$$

where η_n , σ_n , and n are constants. The value of η_n can be chosen arbitrary. In what follows, $\eta_n=10^{-4}\ h^{-1}$, which means that σ_n is the stress to cause 1% creep strain for 100 h. We call this value as creep resistance of a material on 100-hours time base.

A solution of a creep problem for a beam under bending yields a dependence of the deflection rate, $\dot{f}$, of the beam at its centre on applied load Q [43]. For a beam of rectangular cross-section of height $2h$ and width b we have:

$$\dot{f} = \eta_n \frac{1}{2^{3n+2} n^n (n+2)} \left(\frac{Q}{\sigma_n h^2} \right)^n \left(\frac{L}{b} \right)^n \left(\frac{L}{h} \right) L \tag{3}$$

for 3-point bending. The solution was obtained neglecting a contribution of shear deformations to the displacement.

Therefore, measuring $\dot{f}(Q)$ yields the value of exponent n directly and then applying Eq. (3) gives the value of σ_n .

The results of testing some fibres are presented in Figure 21, Figure 22 and Figure 23. In these Figures corresponding data for bulk single crystals are also plotted.

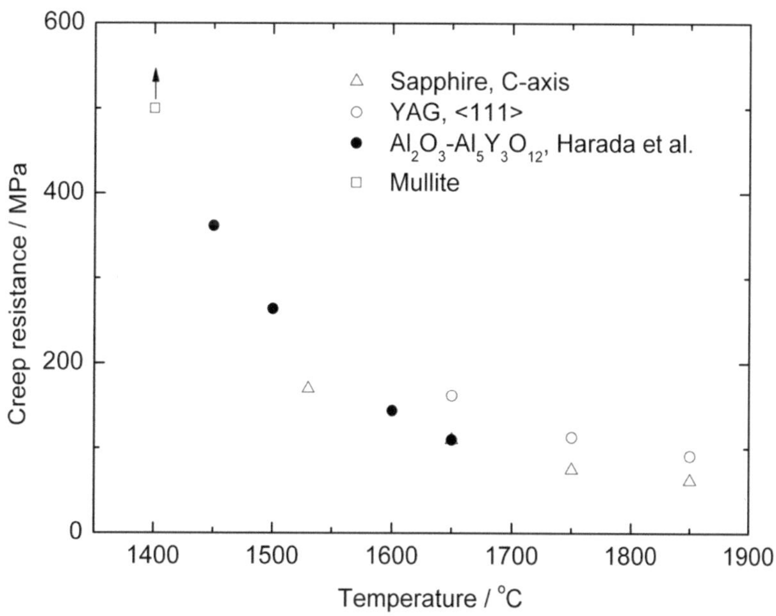

Figure 21. Temperature dependence of the creep resistance (a stress to cause 1% creep strain for 100 h) of some oxides in the form of bulk crystals. Data on sapphire and garnet YAG are after Corman [44], on mullite are after Dokko et al. [35], on Al_2O_3-$Al_5Y_3O_{12}$-eutectics are after Harada et al. [45].

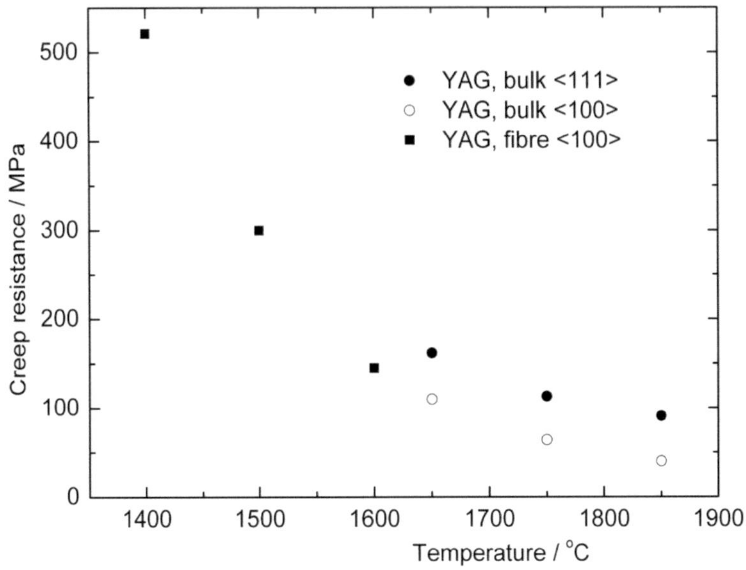

Figure 22. Creep resistance (stress to cause 1% creep strain for 100 h) of bulk YAG crystals of two orientations obtained by using experimental data by Corman [46] and ICM-fibres of YAG obtained by testing oxide/molybdenum specimens loaded in 3-point bending.

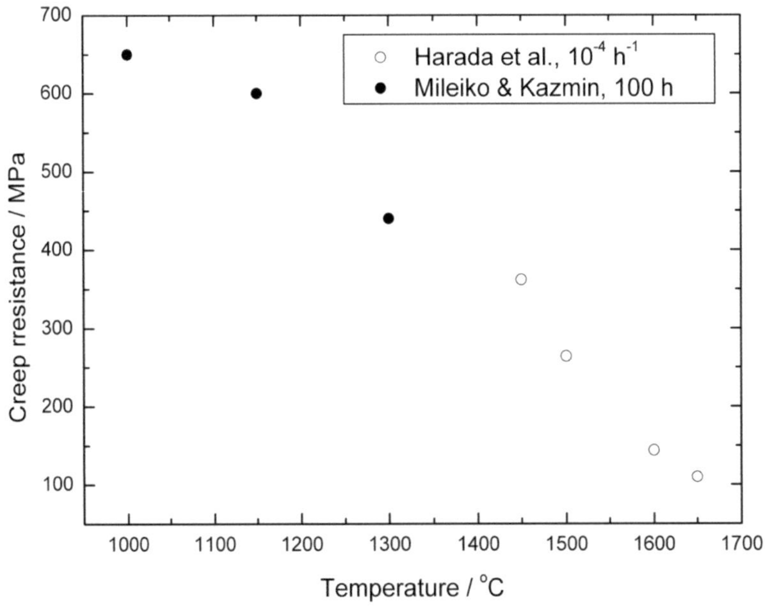

Figure 23. Creep resistance of bulk crystals of alumina-YAG eutectic obtained by using experimental data by Harada et al. [48] and creep rupture stress on the base of 100 h of alumina-YAG eutectic ICM-fibres obtained by using experimental data by Mileiko and Kazmin [46] on tensile creep rupture tests of molybdenum-matrix composites.

Comparing creep resistance of ICM-fibres and bulk crystals for yttrium-aluminium garnet shows an excellent correspondence of creep resistance of bulk crystals and corresponding fibres grown from melt. The data presented provide also an additional confirmation of the validity of the method of evaluation of creep properties of ICM-fibres by testing in bending oxide/molybdenum specimens. Creep resistance of Nextel 720 fibre (α-Al_2O_3+mullite) evaluated from experimental data presented in [50] is also shown in Figure 24. A number of important conclusions can be drawn, in particular:

1. In temperature interval from 1100 to 1600°C, values of the creep resistance of single crystalline YAG and mullite as well as that of alumina-YAG-eutectic fibres obtained by using ICM are nearly the same. YAG fibre looks slightly better than the others. Still, their creep resistance can be certainly enhanced by crystallising them in the <111> direction.

2. Single crystalline mullite fibre produced by ICM does not seem to be superior to, say, YAG fibres. Its creep resistance differs essentially from the experimental point by Dokko et al.

3. Polycrystalline oxide fibres obviously lose their creep resistance below a temperature of 1200°C certainly due to an intrinsic behaviour of grain boundaries (see Figure 24).

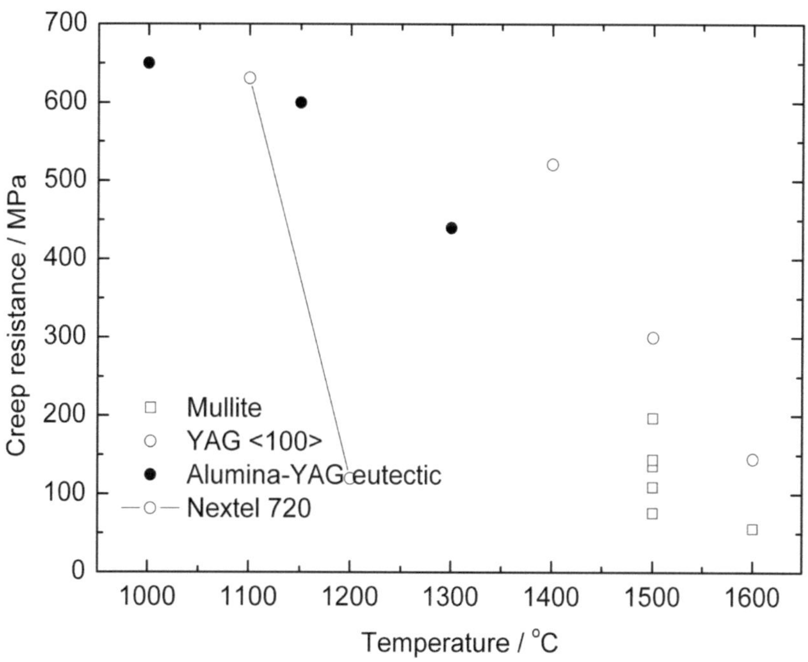

Figure 24. Creep resistance (stress to cause 1% creep strain for 100 h) of ICM-fibres in comparison with commercially available polycrystalline oxide fibres [47].

4. Nickel-based composites reinforced with ICM-fibres

4.1. Preliminary notes

First of all, it should be noted that there are three most important characteristics of heat resistant composites, those being creep resistance, fracture toughness and oxidation resistance. In the case of nickel alloys as matrices for oxide fibres composites sufficiently high oxidation resistance and fracture toughness are to be expected. That is why we will focus here on creep behaviour of nickel-based composites reinforced with ICM-fibres.

On a fundamental study stage of developing composites the main bulk of creep tests are conducted by bending simple composite specimens and then calculating tensile creep characteristics of the composites according to the method developed in [42]. The dependence of displacement rate $\dot{f}$ of the specimen on load Q for a specimen with a circular cross-section of radius R is similar to Eq.(3):

$$\dot{f} = \eta_n \frac{L}{2^{2(n+1)}(n+2)\mu(n)} \left(\frac{Q}{\sigma_n R^2}\right)^n \left(\frac{L}{R}\right)^{n+1} \tag{4}$$

where $\mu(n)$ is expressed via gamma-function $\Gamma(\bullet)$:

$$\mu(n) = \frac{2^{2+\frac{1}{n}}}{3+\frac{1}{n}} \frac{\left(\Gamma\left(1+\frac{1}{2n}\right)\right)^2}{\Gamma\left(2+\frac{1}{n}\right)}.$$

All other symbols are as in Eq. (3).

Such an approach to the creep tests simplifies and speeding up the procedure. The experiments can be even more speeded up if the microstructural creep model developed in reference [51] is used in the analysis of experimental data [48].

The model mentioned yields a dependence of composite stress σ on steady state creep rate $\dot{\varepsilon}$

$$\sigma = \lambda\sigma_m \left[\left(\frac{\sigma_o^{(f)}}{\lambda\sigma_m}\right)^\beta \left(\frac{l_o}{d}\right)\right]^{\frac{m+1}{n}} \left(\frac{\dot{\varepsilon}}{\eta_m}\right)^{\frac{1}{n}} V_f + \sigma_m \left(\frac{\dot{\varepsilon}}{\eta_m}\right)^{\frac{1}{m}} V_m \tag{5}$$

where fibre characteristics are determined by the Weibull based strength/fibre length dependence, $\sigma_*^{(f)}(l) = \sigma_o^{(f)}(l_o)\left(\frac{l}{l_o}\right)^{-\frac{1}{\beta}}$; matrix characteristics are given by the power low of matrix creep,

$\sigma = \sigma_m \left(\frac{\dot{\varepsilon}}{\eta_m} \right)^{\frac{1}{m}}$; $n = m + \beta + m\beta$, d is a characteristic fibre diameter; λ is a function of fibre/matrix interface strength α, $0<\alpha\leq 1$, $\alpha = 1$ correspond to the ideal interface, which means that its strength is equal to the matrix strength.

Let us approximate Eq. (5) by a power function normally accepted while interpreting experimental data on creep of any materials. An important point is a possibility to calculate the value of exponent q in the approximation

$$\sigma = C\dot{\varepsilon}^{1/q} \tag{6}$$

Rewriting Eq. (5) as

$$\sigma = AV_f\dot{\varepsilon}^{1/n} + B(1-V_m)\dot{\varepsilon}^{1/m} \tag{7}$$

and searching for values of q and C to provide the best approximation of Eq. (7) by Eq. (6), which means to find a minimum to the integral

$$\Sigma = \int_{\dot{\varepsilon}_1}^{\dot{\varepsilon}_2} \left(AV_f x^{1/n} + B(1-V_m)x^{1/m} - Cx^{1/q} \right)^2 \mathrm{d}x\,, \tag{8}$$

in which the integration limits are an interval of creep rates of interest. In the present context they are normally 10^{-4} and 10^{-1} h^{-1}.

A comparison of exponent q calculated and the exponent values obtained in the experiments with 2-steps loading of specimens is presented in Figure 25. One can see, that the calculations yield an acceptable result.

Note that a developing of oxide-fiber/nickel-based matrix composites has been meeting a number of the problems. The first one that is finding appropriate oxide fibres has been solved by inventing ICM.

The authors affiliated to NASA described what seemed to be the second problem. They found that sapphire fibres degraded during a liquid infiltration fabrication process. This was observed in testing fibres extracted from the matrix [49]. They concluded that the sapphire fibre is not an appropriate reinforcement for nickel alloys. The experiment was correct, but the conclusion was not completely correct. It will be shown below that the matrix melt interacts with the fibre and causes the fibre degradation but after melt crystallised the matrix heals the defect resulting from the interaction provided there is a good adhesion between the fibre and matrix. Moreover, preexisting defects can also be healed [42, 50]. In fact, the conclusion mentioned can be a kind of the delusion. The real problem is to find ways to organize a strong interface between an oxide fibre and nickel-based matrix.

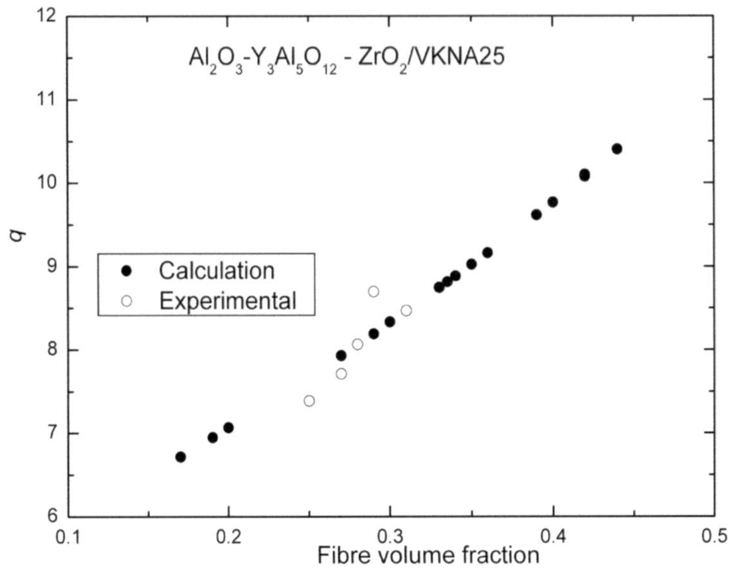

Figure 25. Calculated values of the exponent, q, and the values obtained in the experiments. The values of structural parameters of the composite are: $m = 5$, $\beta = 3$, l_0=1 mm, $d = 0.1$ mm, $\alpha = 0.4$, which corresponds to the effective fibre strength equal to 600 MPa [44]. The test temperature is 1150°C.

4.2. Fabrication of nickel-based composites reinforced with ICM-fibres

Three commercially available Russian superalloys have being used as matrix materials. The alloy marked as VKNA-4U contains Al, Cr, W, Ti, Co, Mo and C. That marked as VKNA-25 differs from VKNA-4U by the presence of 3.5% rhenium. The GS-32 alloy contains Al, Cr, Mo, W, Ta, Co, Nb, Re, C, B.

Since Ni-based alloys do not normally wet oxides, pressure casting is used to produce the composites. To fabricate specimens for mechanical testing, a special simple mould was designed that is composed of external quartz tube with one end closed, internal alumina tube with open ends, porous alumina plug at the top of the fibre bundle, and alumina bed at the bottom of the quartz tube [51]. Temperature and pressure of argon gas in the infiltration process were 1550°C and 5 atm, respectively. Crystallisation of the matrix was performed in the axial temperature gradient to make either a single crystalline or columnar microstructure of the matrix. The diameter and length of the specimens were ~5 and ~50 mm, respectively.

4.3. Mechanical properties of nickel-based composites reinforced with ICM-fibres

Creep experiments were carried in 3-point bending by either step-wise loading and in such case the value of the exponent in a creep power law is determined directly or by a single load, in which case the exponent is calculated as shown above. Then tensile creep characteristics of the composites are calculated according to a procedure also described above. Some examples of dependencies of the creep resistance (accepted as a stress to cause 1% creep strain for 100 h) on fibre volume fraction are presented in Figure 26.

It can be seen that for a particular composite creep resistance can reach a value of 150 MPa at a temperature of 1150°C. It is important to note that the density of the composites under consideration is between 6.5 and 6.8 g/cm^3; the density of modern superalloys are approaching 9 g/cm^3.

A compilation of the data presented in Figure 26 is plotted in Figure 27 as a correlation of the maxima of creep resistance of the composites reinforced with various oxide fibres on fibre volume fraction. Such presentation that shows by arrows the possible developments demonstrates clearly an effect of particular fibre/matrix combinations on the creep properties of the oxide-fibre/nickel-based-matrix composites.

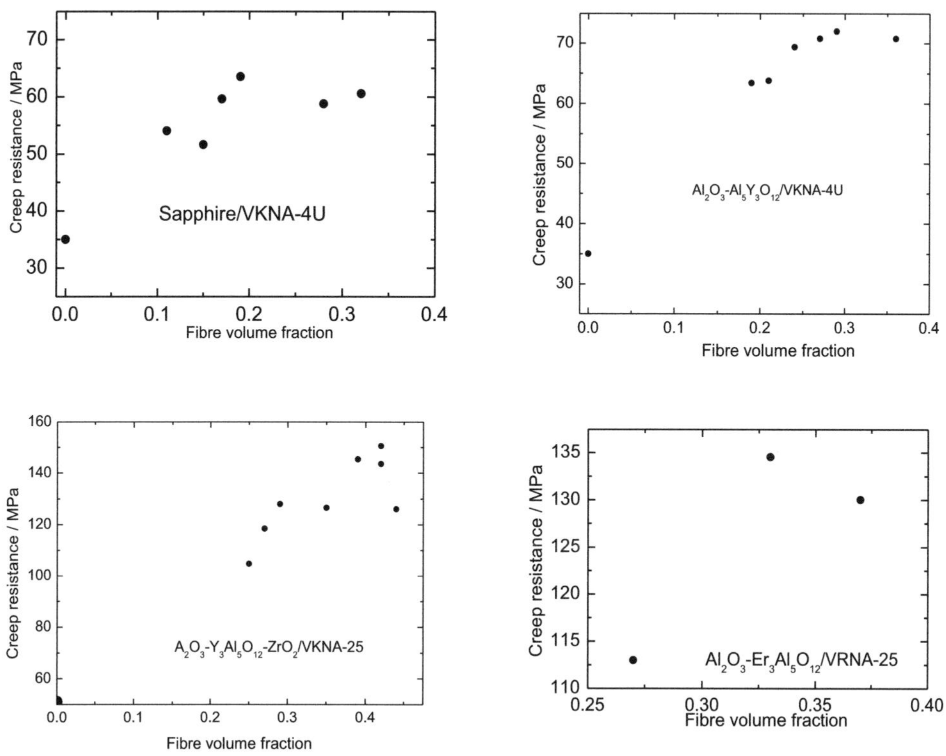

Figure 26. Creep resistance of oxide-fibre/Ni-based-alloy matrix composites versus fibre volume fraction at 1150°C.

Load/displacement curves for AYZ/VKNA-25-matrix composites tested in 3-poit bending at room temperature presented in Figure 28 illustrate two important peculiarities of strength and deformation behaviour of MMCs worth to be mentioned in the present context. First, the curves reveal non-brittle behaviour of the composites even when volume fraction of the brittle constituent reaches 0.48. Such behaviour is to be related to a special fracture toughness properties of MMCs referred to in the Introduction. Secondly, a sequence of the curves for 0 – 0.15 – 0.18 – 0.48 fibre volume fractions reflects another feature of the strength behaviour of

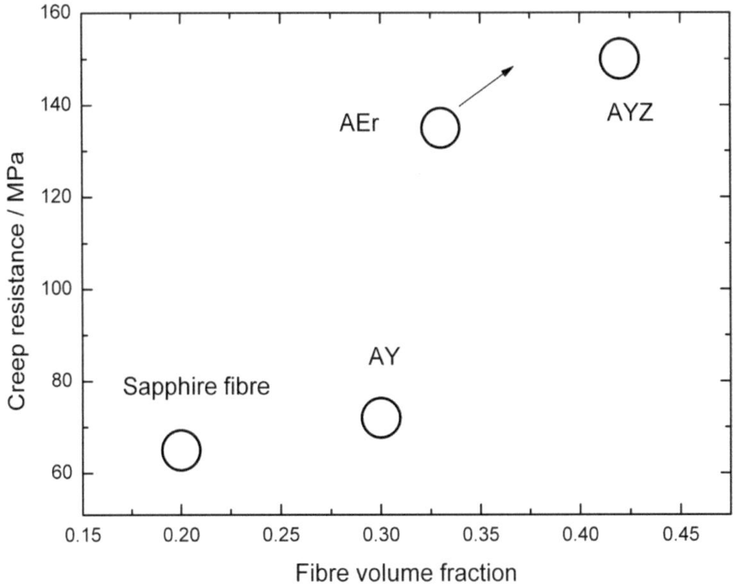

Figure 27. Maxima of the creep resistance of various composites.

MMCs that is the existence of a maximum on a dependence of the strength on fibre volume fraction. This remark is important for a discussion of strength/temperature dependence for an oxide-fibre/high-entropy-matrix composite below, in Section 5.

5. Composites with a high entropy metal matrix

An invention of high entropy alloys (HEAs), which are normally composed of five or more elements each with concentration between 5–35 atomic percent [53, 54] stimulated the opening of a new chapter in the high-temperature metal matrix composites because (i) among such alloys there are some with both high melting points and sufficiently high oxidation resistance [55] and (ii) equiatomic compositions of HEAs promises an easier solution of the wetting problem while dealing with oxide reinforcements. Hence, the author's research group together with a Ukrainian group performed the first experiments aiming at the understanding of prospects of oxide-fibre/HEA-matrix composites [56].

Fibres of Al_2O_3-$Al_5Y_3O_{12}$ and Al_2O_3-$Al_5Y_3O_{12}$-ZrO_2 eutectics were used as reinforcements. A high entropy alloy containing Fe, Co, Ni, Cr and W with a melting point of about 1450°C was chosen as the matrix. The alloy is composed of two phases, those being a base BCC phase and inclusions of a μ-phase of the Fe_7W_6-structure containing all the elements of the alloy, but with their ratios differing from that of the base phase [57].

Specimens were again produced by pressure liquid infiltration of a fibre bundle located in quartz casting mould with the matrix melt. In these experiments, a molybdenum foil 25

http://dx.doi.org/10.5772/61024

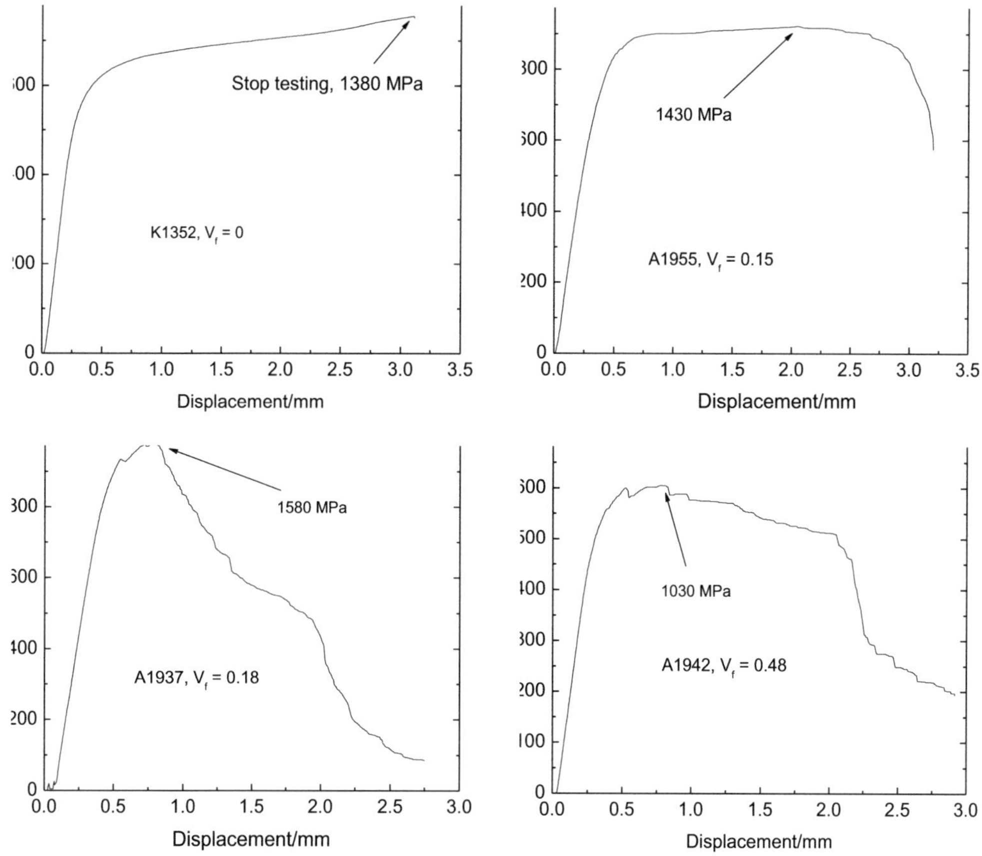

Figure 28. Load/displacement curves for AYZ/VKNA-25-matrix composites tested in 3-poit bending at room temperature [52].

microns thick covered the inside surface of the mould to prevent the interaction between the HEA melt and silica. The necessary gas pressure in this case, 1 – 1.5 atm., is much lower than that in the case of infiltrating oxide fibres with nickel superalloys, which is no less than 5 atm. This is indicative of good wetting in this system, which determines a sufficiently high strength of the fibre/matrix interface.

The microstructure of the composite is illustrated in Figure 29. The results of The X-ray microanalysis of the matrix shows that the base phase (grey phase in the Figure) contains all the elements in nearly equi-atomic ratios, tungsten being an exception. The composition of an inclusion (white phase) is close to $Fe_{1-(x+y+z)}Cr_xCo_yNi_zW_6$. In these experiments, all specimens have relatively large fibre volume fractions, between 40 and 50%.

Composite specimens were tested in the 3-point bending to measure the strength at 20 - 1300°C. Temperature dependence of the strength is presented in Figure 30. The experiments yields two interesting observations.

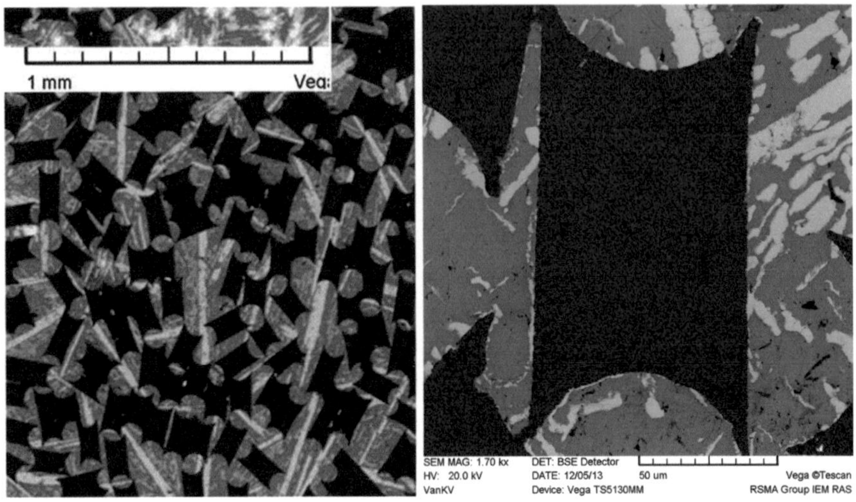

Figure 29. (a) A micrograph of a part of the cross-section of a specimen with Al_2O_3-$Y_3Al_5O_{12}$-eutectic fibre. (b) The microstructure at higher magnification.

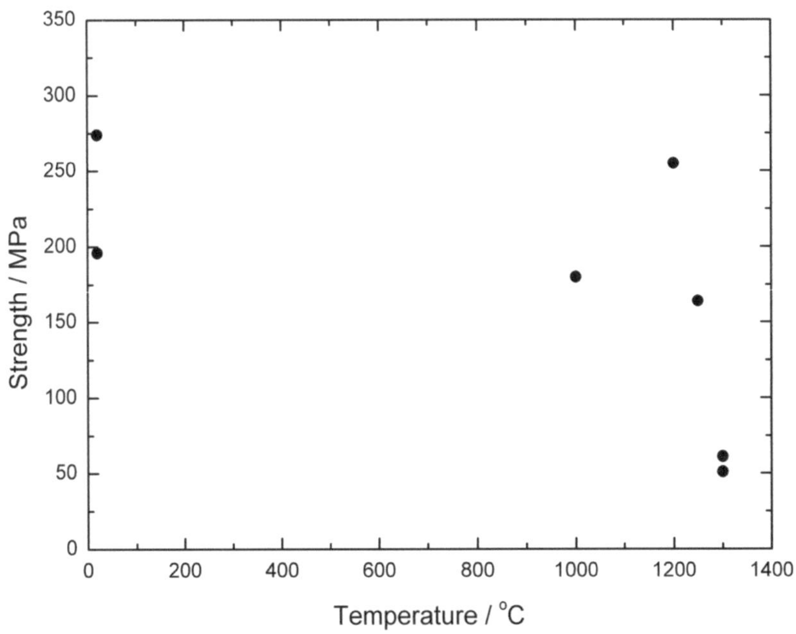

Figure 30. Temperature dependence of the bending strength of the composites reinforced with Al_2O_3-$Y_3Al_5O_{12}$-ZrO_2-eutectic fibre. The fibre volume fractions are between 40 and 50%.

First, the room temperature strength values are relatively low and this is to be the case since the specimens have large fibre volume fractions. The dependence of strength of a brittle-fibre/

http://dx.doi.org/10.5772/61024

ductile-matrix composite on fibre volume fraction V_f has a maximum. An inherent reason for decreasing composite strength at $V_f \geq V_f^{(max)}$ is arising a crack in the composite over a fibre cluster as a result of a single fibre break at a rather low stress followed by fracturing a number of adjacent fibres. The crack size arising because of fracture of a fibre cluster can be too large for the matrix to dissipate the elastic energy and arrest the crack. The probability of fibre cluster formation increases with the fibre volume fraction increases. A corresponding model of the composites failure behaviour was developed by Mileiko [1]. Recent experimental results revealing such behaviour of oxide-fibre/nickel-based matrix composites are published in [56]. A particular value of $V_f^{(max)}$ depends on fracture toughness of the matrix, fibre strength characteristics and a degree of non-homogeneity of fibre packing. Usually, for composites produced by infiltration of a fibre bundle with a matrix melt, $V_f^{(max)} < 0.3$ for room temperature behaviour because of a rather poor homogeneity of the fibre distribution in a specimen cross-section. Hence, the low strength of specimens at room temperature is characteristic of composites with larger fibre volume fraction and the same composites with $V_f < 0.3$ would definitely be characterised by a much higher strength.

Secondly, the strength of composites with $V_f = 0.4 - 0.5$ at room temperature and at 1200°C is approximately the same. Two factors contribute to this effect. First, this can be a result of high fibre/matrix interface strength, which results in loading the fibre to high stresses. The second factor is a shift of critical value V_f(max) of fibre volume fraction to higher values as a result of an increase in the matrix fracture toughness with temperature.

Composites of such a type are to be developed for high temperature use. Hence, creep properties of the composites are of importance. At a prelimary stage of the work described in [60] no creep tests have been performed yet, the results of strength tests together with a previous experience of the present author in studying creep of fibrous composites [51,52, 58] allow formulating some expectations. The theory of creep behaviour of MMCs [51] and a large bulk of the experimental data [56, 59, 60] show that the most important structural parameter of composites with oxide fibres produced via liquid infiltration route is the fibre/matrix interface strength determined by a degree of wetting in the system. If a matrix melt wets a fibre then the matrix heals the surface defects that yields an increase in the effective fibre strength by a factor as high as 5 in the case of a fibre obtained by the internal crystallisation method (see above Figure 13). The effective fibre strength determines creep resistance of composites at temperatures, at which fibre creep is negligible. The oxide fibres used in these experiments do not creep at a temperature of 1200°C (see Section 3). Also it is important to note that good wetting and, consequently, a strong interface make the basis of high creep resistance of the composites with large fibre volume fractions. Therefore, the composites described in short communication [56] can be expected to have a high creep resistance at temperatures to 1200°C.

6. Molybdenum-based composites reinforced with ICM-fibres

As was mentioned in Section 1 the development of refractory metal alloys for high temperature usage meets a problem in reaching an appropriate balance of three main characteristics of such alloys those being creep resistance, fracture toughness (damage tolerance), and oxidation resistance. Hence, the developments of fibrous composites with refractory alloy matrices can obviously be considered as a way to overcome a problem of reaching the balance between creep resistance and fracture toughness. Recent results show a possibility to obtain such composites with an enhanced oxidation resistance [61, 62]. Measurements of oxidation resistance and creep properties were performed on specimens composed of unalloyed molybdenum matrix (containing 0.04% of impurities) and fibers of sapphire, single crystalline YAG and a number of alumina-garnets eutectics (Al_2O_3-$R_3Al_5O_{12}$ where R is Y, Yb or Tb). Three sets of the experiments were performed to evaluate oxidation resistance of the composites.

First, specimens of molybdenum and molybdenum-matrix composites reinforced with alumina-YAG eutectic and sapphire fibers of the same sizes were kept in air at a temperature of 1000°C for up to 1 h. A decrease in the mass of the specimens with time is plotted in Figure 31. A significant difference in the oxidation kinetics of molybdenum reinforced with the fibre containing yttrium compared to both pure molybdenum and molybdenum reinforced with sapphire fibre is obvious. Then the specimen reinforced with the Al_2O_3-$Y_3Al_5O_{12}$ eutectic fibre was exposed to a temperature of 1300°C for nearly 2 h. For comparison, a "fresh" pure molybdenum specimen was also tested. An essential difference in the oxidation of the composite and unreinforced pure molybdenum can be clearly seen. A hypothesis of formation of a sufficiently stable layer of yttrium molybdate on the composite surface attacked by oxygen can explain the result obtained.

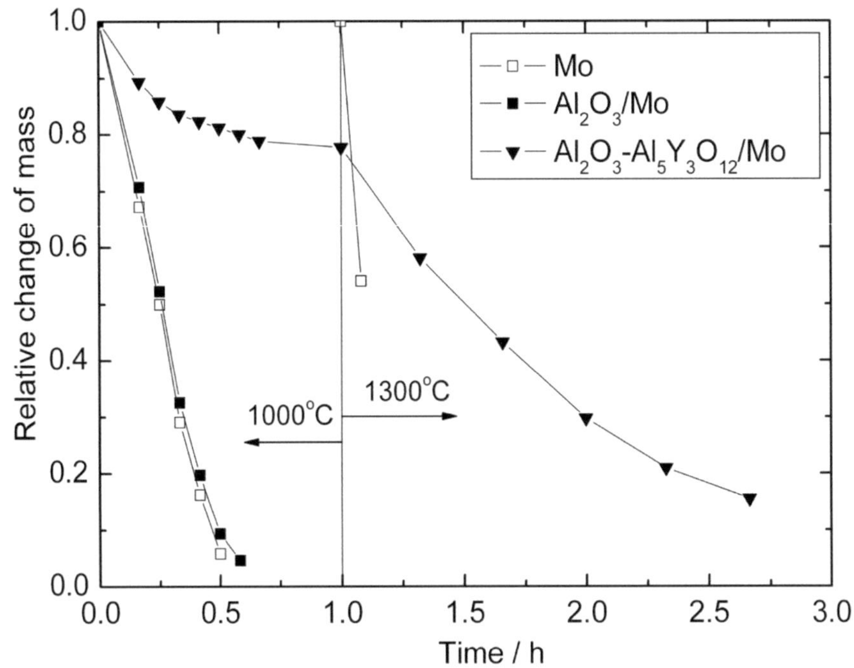

Figure 31. A relative decrease in the mass of molybdenum and oxide/molybdenum specimens with a surface of 20 cm^2 versus time of the exposure in air at 1000 and 1300°C.

The second series of the experiments was carried out to prove the hypothesis just formulated by a direct observation of the surface of a composite with molybdenum matrix and single crystalline YAG fibre heated in air at a temperature of 1000°C for 1 h. In Figure 32, one can see a fibre of a nearly tri-angular shape being originally single crystalline YAG. After heating, a layer in the fibre in a vicinity of the fibre/matrix interface occurs, which looks different from the centre of the fibre. The X-ray microanalysis was performed in three points marked on the micrograph. It was shown that (i) a chemical compound close to yttrium molybdate, $Y_2Mo_3O_{12}$ arises on the surface of both the molybdenum matrix and oxide fibre; (ii) the fibre has lost yttrium oxide and the composition of a layer arising near the interface is close to alumina. Examination of a specimen surface gives exactly the same result (Figure 33): The The X-ray microanalysis performed over areas 60X60 microns around points 4 – 6 shows that the surface is coated with a layer of the yttrium molybdate and depletion of yttria in the fibres is also observed. The only Y_2O_2-MoO_3 phase diagram published in 1982 and cited in [63] shows that the melting point of $Y_2Mo_3O_{12}$ is 1310°C and this sets a limitation for the usage of the composite.

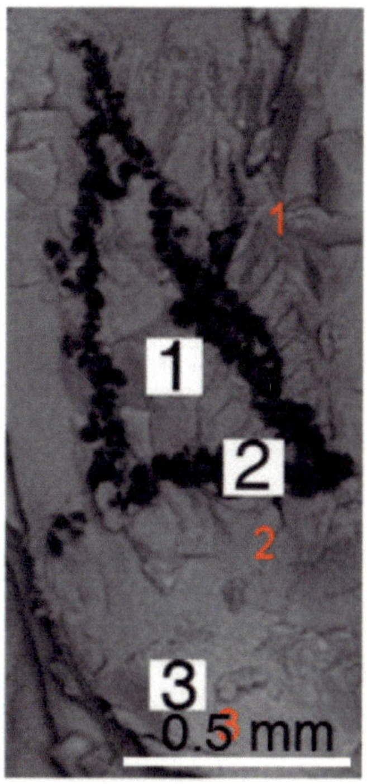

Figure 32. SEM micrograph of a face of a YAG-fibre/molybdenum-matrix specimen heated at 1000°C for 1 h.

Note that the specimens in this experiment had no coating, so a large quantity of yttria was withdrawn from the fibre exposed to a free surface. Obviously, a special coating will prevent molybdenum from oxidation and formation of the molybdate to heal possible defects in the coating will occur only in a vicinity of the damaged area of the coating.

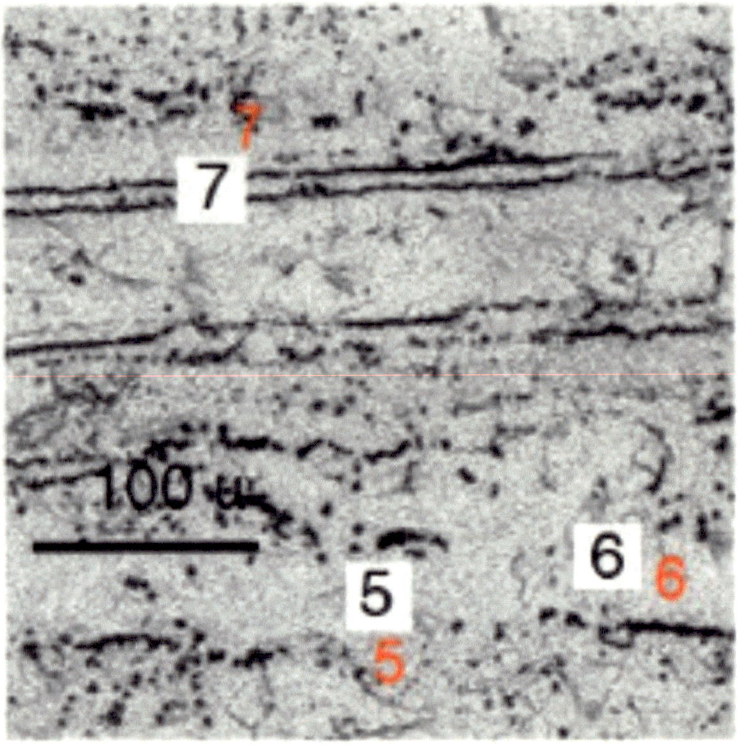

Figure 33. SEM micrograph of the sidepiece of a YAG-fibre/molybdenum-matrix specimen heated at 1000°C for 1 h.

The third set of the experiments was a study of oxidation kinetics of Mo-matrix composites with alumina-based eutectic fibres containing $Y_3Al_5O_{12}$, $Tb_3Al_5O_{12}$ and $Yb_3Al_5O_{12}$. Exposure temperature for these experiments was lower than in the first two sets, 700°C, and exposure time larger, up to 10 h. The oxidation kinetics is presented in Figure 34 together with that for pure molybdenum. The data for molybdenum is taken from reference [64]. It should be noted that those data have been obtained by exposing specimens in atmosphere with at a partial oxygen pressure of 76 mm Hg, which is much lower than that in normal air atmosphere. Still, the oxidation rates of the composites are much smaller.

A layer of a compound containing both molybdenum and yttrium is found on the surface of the specimen with Al_2O_3-$Y_3Al_5O_{12}$ eutectic fibres with atomic ratio Mo:Y ≈ 2:1, which corresponds to $Y_2Mo_4O_{15}$. Comparing this observation with the data obtained in the second set of the experiments leads to a hypothesis on the formation of a sequence of the molybdates with enhancing the exposure temperature. Nearly the same is found on the surface of a composite with eutectic fibers containing lanthanide elements. Again the Mo:Tb ratio is approximately 2. In the absence of published phase diagrams of a corresponding oxide system, it remains to expect an occurrence of $Tb_2Mo_4O_{15}$ on the specimen surface.

Tensile creep tests of the composites with Al_2O_3-$Y_3Al_5O_{12}$-eutectic fibre as model materials were conducted in vacuum at temperatures 1000 - 1300°C [27]. The dependence of rupture stress on

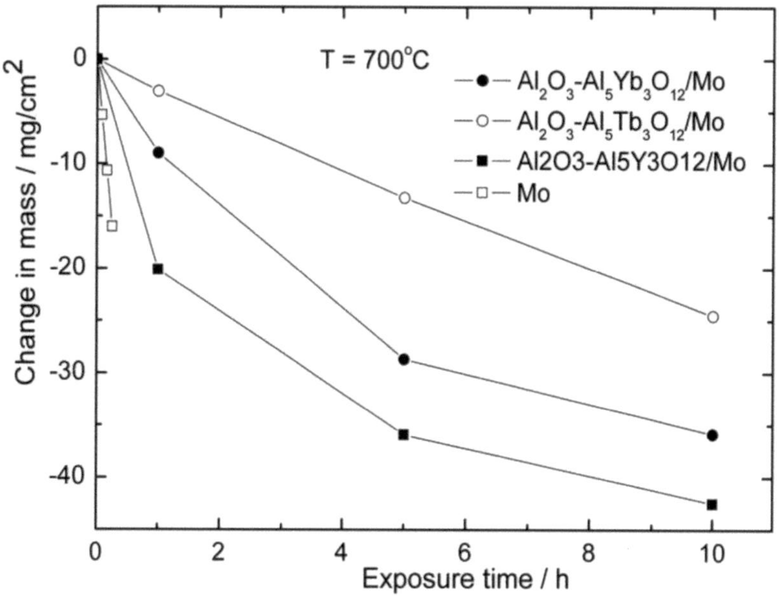

Figure 34. Oxidation kinetics at 700°C of three Mo-matrix composites with oxide fibers containing rare earth elements and that of unreinforced molybdenum. Experimental points for composites containing Al_2O_3-$Tb_3Al_5O_{12}$ and Al_2O_3-$Yb_3Al_5O_{12}$ eutectic fibres are the averages for three specimens. Those for composites containing Al_2O_3-$Y_3Al_5O_{12}$ fibres are the averages for six specimens.

time for temperature 1300°C is presented in Figure 35 together with the data acquired from experimental results published by Jain and Kumar [11], which describe creep rupture behaviour of molybdenum alloy of sufficiently high oxidation resistance. Creep rupture properties of the composite with a low density (~ 7.5 g/cm^3) on 100 h time base are better than those of the alloy. At lager rupture time the difference between values of the rupture stress for the composite and alloy will be even larger. This is because creep resistance of metal matrix composites with fibres of creep resistance much higher that of the matrix is determined by the fibre strength (see Eq. (5)) that yields a slop of the stress/time dependence in log-log coordinates for the composite much smaller than that for an alloy.

7. Prospects of the internal crystallisation method

The internal crystallisation method can be further developed to produce fibres and composites differ from those described in the previous sections. We limit ourselves just with two examples to illustrate the work in progress.

7.1. Composite oxide fibres

Fibrous composites containing inherently brittle matrix and brittle fibre can behave in a non-brittle manner provided a fibre/matrix interphase allows either fibre/matrix delamination or

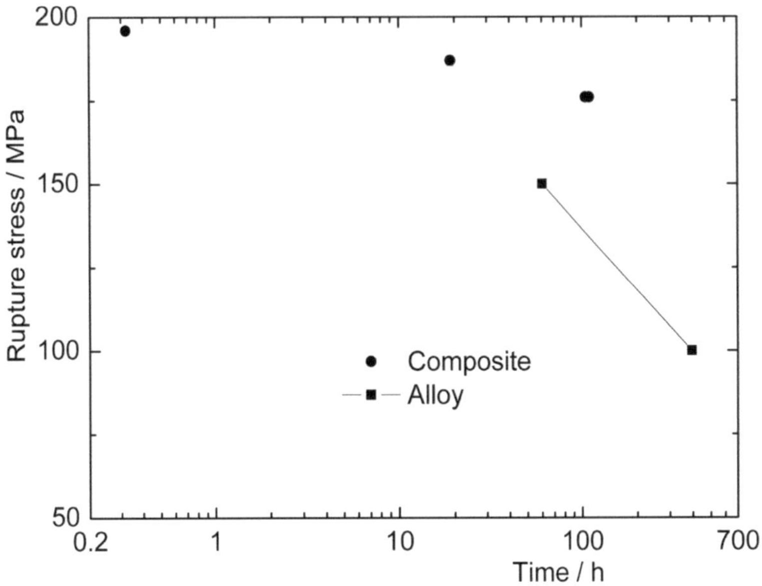

Figure 35. Creep strength of an Al_2O_3-$Y_3Al_5O_{12}$/Mo composite with fibre volume fraction of about 40% and that of most creep resistant three-phase Mo-Si-B alloy (acquired from experimental results in [11])

fibre pull-out. A usual way to organise such a kind of the "weak" interphase is to coat the fibre with an appropriate material. In the case of oxide-fibre/oxide-matrix composites, polycrystalline fibres, which are commercially available, are coated with either monazite or hexaaluminate type of the oxides [65, 66, 67]. Hexaaluminate $CaAl_{12}O_{19}$ (CA6 further on) has a hexagonal highly anisotropic structure, so that fracture-energy anisotropy reaches values of about 100 [71]. However, polycrystalline oxide fibres are creeping at rather low temperatures (see Figure 24), so the usage of single crystalline or eutectic fibres of the ICM type is wanted. Coating of ICM fibres is possible [68] but not a convenient procedure.

Therefore, a new design of the fibrous microstructure able to arrest cracks should be looking for. A possible solution is a composite oxide fibre containing a main ingredient, say sapphire, and a "weak" inclusion, say a kind of hexaaluminate, obtained by using the internal crystallization method [69]. To crystallise such fibres, raw mixtures of oxides containing 2 – 3 – 4.5 – 6 mass per cent of calcia in alumina were prepared to be used in the ICM process. Pulling rate of the oxide/molybdenum blocks varied from less than 1 mm/min to 800 mm/min. In the first experiment, just to show a possible microstructure of hexaaluminate, pure CA6, without alumina was crystallised, the result is presented in Figure 36. It can be seen that the flat surface of the fibre is coincided with the basis crystallographic plane of CA6. Also it is obvious that the fibre is not single crystalline and the fibres of such type cannot be sufficiently strong.

Composites with Al_2O_3-$CaAl_{12}O_{19}$ fibre and brittle molybdenum matrix (molybdenum in a block shown in Figure 3 is brittle being fully recrystallised) demonstrate quasi-plastic behaviour as can be seen in Figure 37a. Fracture surface (Figure 37b) of a composite specimen of the same type reveals fibre pull out.

http://dx.doi.org/10.5772/61024

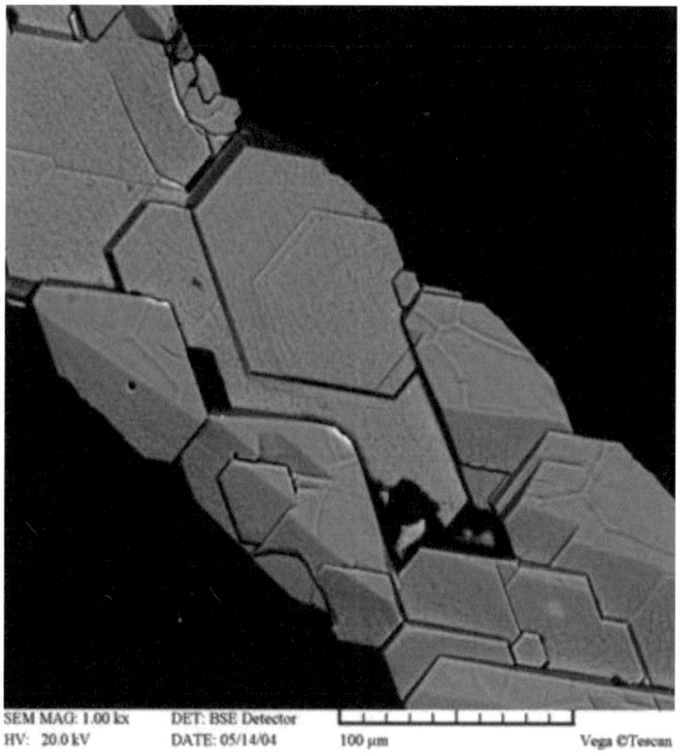

Figure 36. A flat surface of the fibre made of pure hexaaluminate.

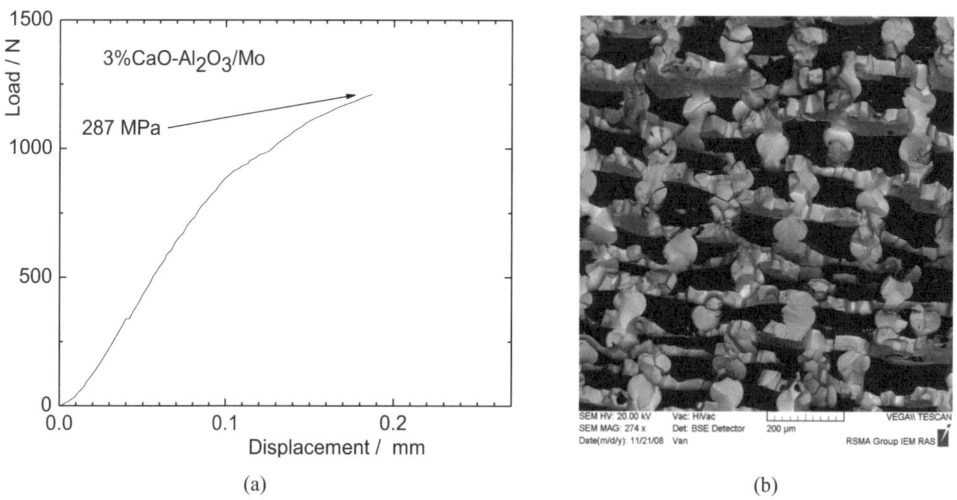

Figure 37. Load/displacement curve of a composite with brittle molybdenum matrix under 3-point bending (a); Failure surface of similar specimen with a fibre obtained from raw mixture Al_2O_3+2%CaO at pulling rate of 1 mm/min.

Hence, there is a hope to develop composite oxide fibres to be an effective reinforcement for brittle matrix.

7.2. Silicide-fibre/molybdenum-matrix composites

As shown in Section 1 a mainstream in the development of heat resistant materials beyond nickel superalloys is searching for molybdenum alloy with a necessary balance of three main characteristics, those being creep resistance, damage tolerance and oxidation resistance. A goal to develop an alloy with the use temperature as high as 1300°C is expected to be reached by introducing into molybdenum matrix brittle particles of complex silicides. However, the alloys with sufficiently high creep and oxidation properties remain to be optimized to satisfy damage tolerance requirement. At the same time, it was mentioned in Section 1 that reinforcing a ductile metal matrix with brittle fibre can lead to a composite of an enhanced fracture toughness properties. Hence, it is interesting to look at prospects of the internal crystallization method to make silicide-fibre/refractory-metal-matrix composites. Preliminary experiments with silicide-fibre/molybdenum-matrix composites were conducted recently [70].

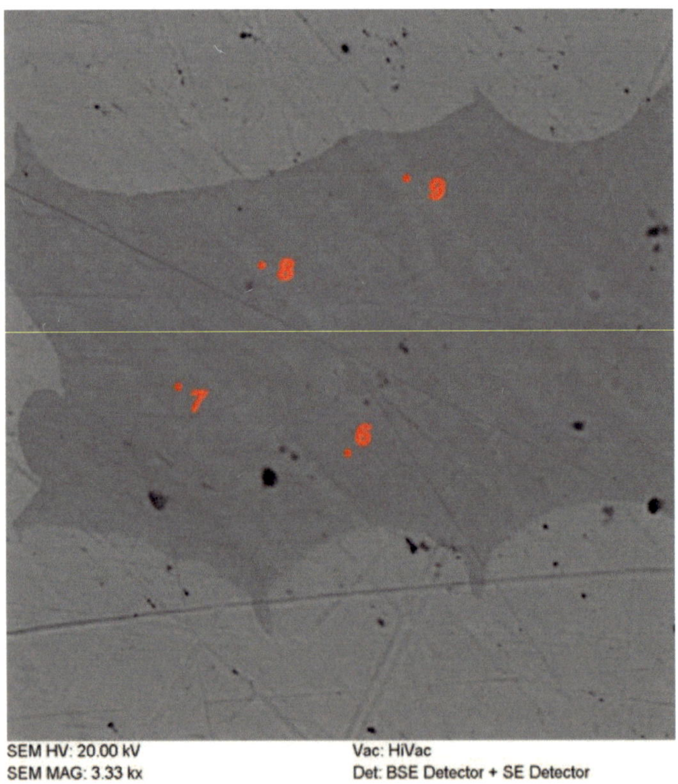

Figure 38. A typical microstructure of the molybdenum-silicide-fibre/molybdenum-matrix composite.

A most difficult couple of the ingredients to deal with is a molybdenum carcass and molybdenum disilicide, $MoSi_2$, with the melting temperature 2020°C and this couple was chosen for the first experiments. The $MoSi_2$ melt was expected to interact with molybdenum to form silicides with a larger content of Mo, Mo_5Si_3 or even Mo_3Si. In fact, the $MoSi_2$ -melt/molybdenum interaction yields a drastic change in the fibre shape and its composition. The fibre lost its plane surfaces, Figure 38, they show traces of a complicated process of the chemical interactions. The Mo:Si atomic ratio measured in the points marked in the area of the fibre is about 3.4, which is close to silicide Mo_3Si. The The X-ray microanalysis of the matrix reveals a presence of silicon, certainly in the form of inclusions of Mo_3Si of small size since the The X-ray spectrum obtained from a cross-section of a specimen (Figure 39) shows only molybdenum and Mo_3Si.

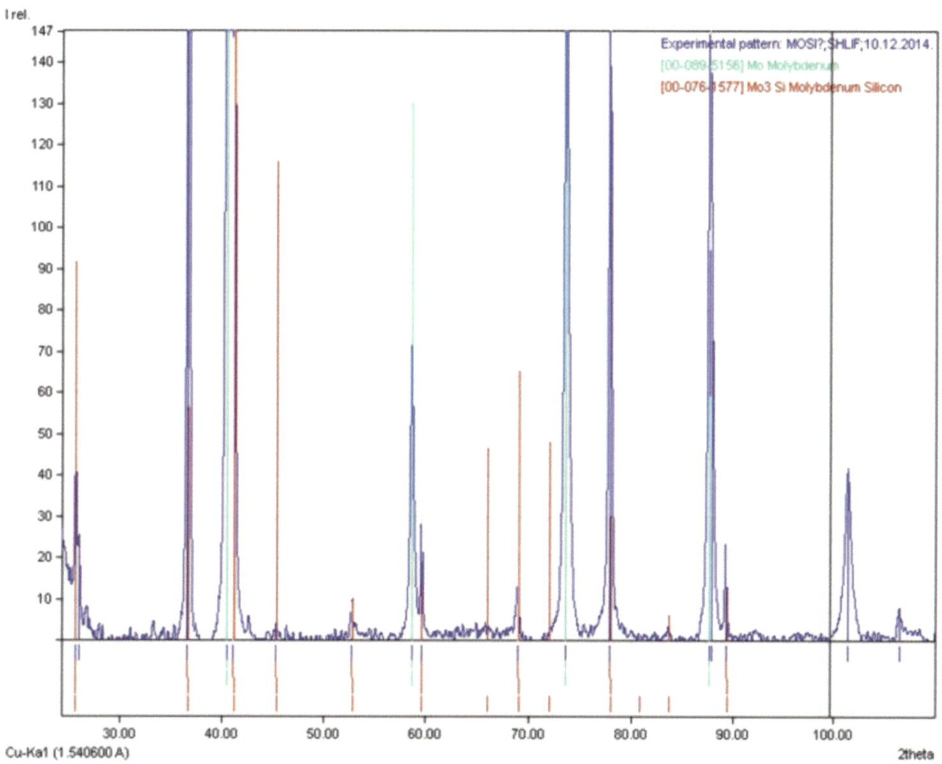

Figure 39. The X-ray spectrum obtained from a cross-section section of the molybdenum-silicide-fibre/molybdenum-matrix specimen.

Creep tests of specimens were conducted in bending at a temperature of 1410°C and the results of the experiments were interpreted on the basis of the method outlined above. One specimen was tested under a step-wise changing of the load, and the value of n occurred to be around 5. All the data were obtained by assuming $n = 5$; they are presented in Figure 40.

It is important to note that value of $n = 5$ was evaluated for a composite with fibre volume fraction $V_f = 0.2$. Because the value of n increases with V_f increasing (see Figure 25, in which the exponent was denoted as q), the value of creep resistance for composites with V_f different from 0.2, which were calculated assuming $n = 5$, are to be corrected as shown in Figure 40 by the arrows. This would make the dependence of the creep resistance for fibre volume fraction close to linear, which is characteristic for composites with a strong fibre/matrix interface.

Therefore the preliminary data just described show that the ICM is a prospective way to obtain silicide fibrous composites.

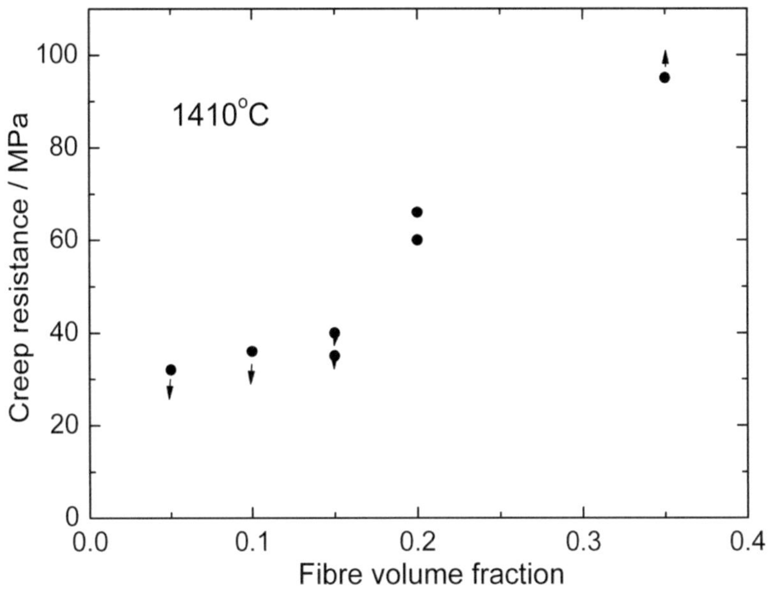

Figure 40. Creep resistance at 1410°C of Mo_3Si-Mo composites versus fibre volume fraction.

8. Conclusions

1. The internal crystallisation method (ICM) can be a base for developing effective technologies to produce a large family of oxide fibres with mechanical properties similar to those of the fibres produced by using traditional methods.

2. ICM-fibres can be used to obtain heat resistant composites with a number of the matrices including nickel superalloys and recently invented high entropy alloys.

3. An appropriate choice of the chemical composition of oxide reinforcements for refractory matrices can be a way to reach a necessary balance of the main properties of heat resistant materials, those being creep resistance, fracture toughness and oxidation resistance.

4. Composite ICM-fibres containing highly anisotropic oxide inclusions can be an effective reinforcements for brittle matrices to produce non-brittle composites.

5. Silicide fibrous reinforcements can also be produced by using ICM.

Acknowledgements

New results reported in the present chapter have been obtained in the framework of projects supported by Russian Foundation for Basic Research (## 13-08-90459, 14-08-01254 and 15-03-05415). The author is thankful to his colleagues who worked hard even in the difficult for Russian science times, years 1990-2005. Special thanks to Drs A.V. Serebryakov, V.I. Kazmin, A.A. Khvostunkov, V.V. Tvardovsky, N.I. Novokhatskaya, A. M. Rudnev and A.N. Tolstun, Mrs. N.A. Prokopenko, Mr. L.S. Kozhevnikov, Mr. A.Ya. Mizkevich, Mr. A.A. Kolchin, Mr. V.A. Chumichev, Mr. D.G. Pizhenin. Regrettably, some of them (AVS, VIK, AAKh and LSK) have passed away..

Author details

Sergei T. Mileiko

Address all correspondence to: mileiko@issp.ac.ru

Institute of Solid State Physics of RAS, Moscow, Russia

References

[1] Mileiko ST. Metal and Ceramic Based Composites. Amsterdam: Elsevier; 1997. 702 p. ISBN: 978-0-444-82814-9

[2] Logunov AV, ShmotinYuN. Advanced heat resistant nickel alloys for gas turbine disks. Nauka I Techologii: Moscow; 2013. 256 p. (in Russian).

[3] Naslain R. Fibrous ceramic-ceramic composite materials processing and properties.Journal de Physique. 1986;47:C1-703-715. DOI: 10.1051/jphyscol:19861107

[4] GE9X - GE Aviation. Available from: http://www.geaviation.com/newengine/ Accessed: 2015-06-02

[5] Lange FF, Tu WC, Evans AG.Processing of damage-tolerant, oxidation-resistant ceramic matrix composites by a precursor infiltration and pyrolysis method. Mater Sci-Eng A. 1995;195:145–150. DOI:10.1016/0921-5093(94)06513-6

[6] Zok FW.Developments in oxide fiber composites. J Am Ceram Soc. 2006;89:3309–3324. DOI: 10.1111/j.1551-2916.2006.01342.x

[7] Volkmann E, Tushtev K, Koch D, Wilhelmi C, Göring J, Rezwan K. Assessment of three oxide/oxide ceramic matrix composites: Mechanical performance and effects of heat treatments. Composites Part A. 2015;68:19–28. DOI: 10.1016/j.compositesa.2014.09.013

[8] Grammenos I, Tsakiropoulos P. Study of the role of Hf, Mo and W additions in the microstructure of Nb-20Si silicide based alloys. Intermetallics. 2011;19: 1612-1621. DOI:10.1016/j.intermet.2011.06.008

[9] US Patent 7,704,335 published April 27, 2010.

[10] Tanaka R, Kasama A, Fujikura M, Iwanaga I, Tanaka H, Matsumura Y. Research and development of niobium-based superalloys for hot components of gas turbines. In: Proceedings of International Gas Turbine Congress; 2-7 November 2003; Tokyo: p.1-5

[11] Jain V, Kumar K S. Tensile creep of Mo–Si–B alloys.Acta Mater.2010; 58:2124–2142.DOI:10.1016/j.actamat.2009.11.054

[12] Heilmaier M, Krüger M, Saage H, Rösler J, Mukherji D, Glatzel U, Völkl R, Hüttner R, Eggeler G, SomsenCh, Depka T. Metallic materials for structural applications beyond nickel-based superalloys. J. Metals.2009; 61:61-67. DOI: 10.1007/s11837-009-0106-7

[13] Lemberg J .A, Middlemas M R, Weingärtner T, Gludovatz B, Cochran J K, Ritchie R O. On the fracture toughness of fine-grained Mo-3Si-1B (wt.%) alloys at ambient to elevated (1300oC) temperatures. Intermetallics. 2012; 20;141-154. DOI:10.1016/j.intermet.2011.09.003

[14] LaBelle H E, Jr, Mlavsky A I. Growth of sapphire filaments from the melt. Nature; 1967; 216:574-575. DOI:10.1038/216574b0

[15] Stepanov AV.A Future of Metal Processing. Leningrad: Machinostroenie; 1963 (in Russian).

[16] Mileiko S T. Oxide fibres. In: Watt WW and Perov BV, eds. Strong Fibres, Handbook of Composites. Vol. 1. Amsterdam: North-Holland; 1985. p. 87- 114.ISBN-13: 978-0444875051 ISBN-10: 0444875050

[17] Fukuda T, Chani V I, editors, Shaped Crystals Growth by Micro-Pulling-Down Technique. Berlin Heidelberg: Springer-Verlag; 2007. ISBN: 3540712941

[18] Burrus C A, Coldren L A. Growth of single-crystal sapphire-clad ruby fibers.ApplPhysLett 1977; 31:383-384.DOI:org/10.1063/1.89713

[19] Feigelson S H, Kway W L, Route R K. Single crystal fibers by the laser-heated pedestal growth method. Optical Engineering.1985; 24:822-830. doi:10.1117/12.7973637

[20] Haggerty J S, Wills K S, Sheehan J E. Growth and properties of single crystal oxide fibers. In: Proceedings of the 15th Annual Conference on Composites and Advanced Ceramic Materials, Part 2 of 2: Ceramic Engineering and Science Proceedings, Volume 12, Issue 9/10; 1991; p. 1785-1801 DOI: 10.1002/9780470313848.ch13

[21] Shahinian P. High-temperature strength of sapphire filaments. J Amer Ceram Soc. 1971;54:67-68. DOI:10.1111/j.1151-2916.1971.tb12180.x

[22] Hurley GF.Short-term elevated temperature tensile behaviour in 0o sapphire filaments. J Mater Sci 1972;7:471-473.

[23] Kotchick DM, Hink RC, Tressler RE. Gauge length and surface damage effects on the strength distributions of silicon carbide and sapphire filaments. J Compos Mater 1975; 9:327-340.

[24] Newcomb SA, Tressler RE.High-temperature fracture toughness of sapphire, J Am Ceram Soc. 1994; 77:3030-3032.DOI: 10.1111/j.1151-2916.1994.tb04542.x

[25] Mileiko ST, Kazmin VI. Crystallization of fibres inside a matrix: a new way of fabrication of composites. J Mater Sci. 1992; 27: 2165-2172.DOI: 10.1007/BF01117932

[26] Mileiko ST, Kazmin VI. Structure and mechanical properties of oxide fibre reinforced metal matrix composites produced by the internal crystallization method. CompSci Tech. 1992; 45:209-220. DOI:10.1016/0266-3538(92)90081-D

[27] KurlovVN, KiikoVM, Kolchin AA, Mileiko ST. Sapphire fibres grown by a modified internal crystallization method.J Crystal Growth.1999; 204:499-504.DOI: 10.1016/S0022-0248(99)00213-4

[28] Mileiko ST, Kiiko VM, Sarkissyan NS, StarostinMYu, Gvozdeva SI, Kolchin AA, Strukova GK. Microstructure and properties of Al2O3-Al5Y3O12 fibre produced via internal crystallization route. Compos Sci Tech. 1999; 59:1763-1772; DOI: 10.1016/S0022-0248(99)00213-4

[29] Mileiko ST, Kurlov VN, Kolchin AA, Kiiko VM.Fabrication, properties and usage of single-crystalline YAG fibres.Journal of the European Ceramic Society.2002; 22:1831-1837.DOI:10.1016/S0955-2219(01)00505-2

[30] Rüscher CH, Mileiko ST, Schneider H. Mullite single cyrystalfibers produced by the internal crystallization method (ICM). Journal of the European Ceramic Society.2003; 23: 3113–3117.DOI:10.1016/S0955-2219(03)00118-3

[31] Mileiko ST. Single crystalline oxide fibres for heat-resistant composites. Compos.Sci. and Technol. 2005; 65: 2500-2513.DOI: 10.1016/j.compscitech.2005.05.029

[32] Mileiko ST, Serebryakov AV, Kiiko VM, Kolchin AA, Kurlov VN, Novokhatskaya NI,Tolstun AN. Single crystalline mullite fibres obtained by the internal crystallisation method. Composites and Nanostructures, 2009; #2: 47-46 (in Russian).

[33] Mileiko ST, Serebryakov AV, Kiiko VM, Kolchin AA, Kurlov VN, Novokhatskaya NI. Single crystalline mullite fibres obtained by the internal crystallisation method:

microstructure and creep resistance. J Europ Ceram Soc. 2009; 29: 337-345.DOI: 10.1016/j.jeurceramsoc.2008.06.022

[34] Mileiko ST, Kiiko VM, Novokhatskaya NI,Tolstun AN, Kolchin AA. Single crystalline fibres of yttrium-aluminium garnet: microstructure and room temperature strength. Deformation and Fracture of Materials.2008; #5: 2-7 (in Russian).

[35] Dokko PC, Pask JA, Mazdiyasni KS.High-temperature mechanical properties of mullite under compression. J. Am. Ceram. Soc. 1977; 60: 150-155DOI: 10.1111/j. 1151-2916.1977.tb15492.x

[36] Kriven WM, Palko JW, Sinogeikin S, Bass JD, Sayir A, Brunauer G, Boysen H, Frey F, Schneider J. High temperature single crystal properties of mullite, J European Ceramic Society.1999;19:2529-2541.DOI: 10.1016/S0955-2219(99)00124-7

[37] Pask JA.Importance of starting materials on reactions and phase equilibria in the AI2O3-Si02 system. J. European Ceram. Soc. 1996; 16: 101-108.DOI: 10.1016/0955-2219(95)00147-6

[38] Mileiko ST, Kiiko VM, StarostinM.Yu, Kolchin AA, Kozhevnikov L.S. Fabrication and some properties of single crystalline mullitefibers. ScriptaMaterialia. 2002; 22:1831–1837.DOI: 10.1016/S0955-2219(01)00505-2

[39] Tolstun AN, Kiiko VM, Kolchin AA, Kurlov VN, Novokhatskaya NI, Mileiko ST. Fabrication, microstructure and high temperature mechanical propertyies of some oxide eutectic fibres. Deformation and Fracture of Materials. 2007; #3: 12-20 (in Russian).

[40] Mileiko ST, Sarkissyan NS, Kolchin AA, Kiiko VM. Oxide fibres in a Ni-based matrix – do they degrade or become stronger? Journal of Materials:Design and Applications. 2004; 218:193-200.DOI: 10.1177/146442070421800303

[41] Kiiko VM, Mileiko ST. Evaluation of room temperature strength of oxide fibres produced by the internal crystallization method.Compos.Sci. and Technol. 1999; 59:1977-1981.DOI:10.1016/S0266-3538(99)00054-8

[42] Mileiko ST. Oxide-fibre/Ni-based matrix composites – III: A creep model and analysis of experimental data.Compos Sci Tech. 2002; 62:195-204.DOI: 10.1016/ S0266-3538(01)00162-2

[43] RabotnovYuN, Mileiko ST. Short-Time Creep. Moscow: Nauka Publishers. 1970 224 p. (in Russian).

[44] Corman GS.High-temperature creep of some crystal oxides. Proceedings of the 15th Annual Conference on Composites and Advanced Ceramic Materials, Part 2 of 2: Ceramic Engineering and Science Proceedings, Volume 12, Issue 9/10: p. 1745-1766. DOI: 10.1002/9780470313848.ch10

[45] Harada Y, Suzuki T, Hirano K, Waku Y.Ultra-high temperature compressive creep behavior of an in-situ Al2O3 single-crystal/YAG eutectic composite. JEuropeanCeramSoc. 2004; 24:2215-2222.DOI:10.1016/S0955-2219(03)00640-X

[46] Mileiko ST, Kazmin VI.Fabrication of composites by the internal crystallization method.MechanicaCompositnychMaterialov. 1991; #5:898-908 (in Russian).

[47] Wilson DM, Visser LR.High performance oxide fibers for metal and ceramic composites.Composites Part A:2001; 32:1143–1153. DOI:10.1016/S1359-835X(00)00176-7

[48] Mileiko ST. Speeding up creep tests of novel high temperature composites. In: Proceedings of 15th International Conference on Experimental Mechanics (ICEM-15). 22-27 July 2012; Porto: INEGI; 2012. P. 13-14.

[49] AsthanaR, Tewari SN, Draper SL. Strength degradation of sapphire fibers during pressure casting of a sapphire-reinforced Ni-base superalloy.Metallurgical and Materials Transactions A: Physical Metallurgy and MaterialsScience. 1998; 29A:1527-1530. DOI: 10.1007/s11661-998-0370-0

[50] Asthana R, Mileiko ST, Sobczak N. Wettability and interface considerations in advanced heat-resistant Ni-based composites. Bulletin of the Polish Academy of Sciences, Technical Sciences.2006;54:147-166.

[51] Mileiko ST, Kiiko VM, Kolchin AA, Serebryakov AV, Korzhov VP, Starostin My, Sarkissyan NS. Oxide-fibre/Ni-based matrix composites – I: Fabrication and microstructure, Compos. Sci. and Technol. 2002; 62:167-179.DOI:10.1016/S0266-3538(01)00197-X

[52] Novokhatskaya NI, Tolstun AN, Kiiko VM, Kolchin AA, Mileiko ST.An effect of nonhomogeneous fibre packing on mechanical properties of oxide/nickel composites.Composites and Nanostructures, 2011; #1:5-17 (in Russian).

[53] Yeh JW, Chen SK, Lin SJ, Gan JY, Chin TS, Shun TT, Tsau CH, Chang SY. Nanostructured high-entropy alloys with multiple principal elements: Novel alloy design concepts and outcomes. Adv Eng Mater 2004; 6:299–303. DOI: 10.1002/adem.200300567

[54] Zhang Y, Zhou YJ.Solid Solution Formation Criteria for High Entropy Alloys.Materials Science Forum. 2007; 561-565:1337-1339. DOI: 10.4028/www.scientific.net/MSF.561-565.1337

[55] Liu CM, Wang HM, Zhang SQ, Tang HB, Zhang AL. Microstructure and oxidation behaviour of new refractory high entropy alloys. J Alloys Compounds 2014; 583:162–169. DOI: 10.1016/j.jallcom.2013.08.102

[56] Mileiko ST, Firstov SA, Novokhatskaya NI, Gorban' VF, Krapivka NP.Oxide-fibre/high-entropy-alloy-matrix composites. Composites Part A 2015;76:131-134. 10.1016/j.compositesa.2015.05.023

[57] Firstov SA, Gorban VF, Krapivka NA, Pechkovsky EP, Samilyuk AV. High temperature mechanical properties of cast nano-structured high entropy alloys. In: Proceed-

ings of 4th Intern Conf Deformation and Fracture of Materials and Nanomaterials, Moscow: IMET RAS, 2011 p 396-398.

[58] Mileiko ST. Steady state creep of a composite material with short fibres.1970; 5:254-261. DOI: 10.1007/BF00551002

[59] Mileiko ST, Kiiko VM, Kolchin AA, Korzhov VP, Prokopenko VM, Oxide-fibre/Ni-based matrix composites – II: Mechanical behaviour. Compos.Sci. and Technol. 2002, 62:181-193. DOI: 10.1016/S0266-3538(01)00161-0

[60] Mileiko ST, Kiiko VM, Kolchin AA, Novokhatskaya NI, Van KV, Bazyleva OA, BondarenkoYu.A. Creep of oxide/nickel composites. Composites and Nanostructures. 2009; №4:5-18 (in Russian).

[61] Mileiko ST, Novokhatskaya NI.On a possibility to make heat resistant composites of high gas corrosion resistance based on refractory metal matrix. Composites and Nanostructures.2012; # 4:5-14 (in Russian).

[62] Mileiko ST, Novokhatskaya NI.High temperature oxide-fibre/molybdenum-matrix composites of improved oxidation resistance, Journal of Materials Engineering and Performance.DOI: 10.1007/s11665-014-1305-0

[63] Phase Equilibria Diagrams, Acer – NIST, Version 3, CD-ROM Database (2003).

[64] MorgunovaNI, Klypin BA, Boyarshinov VA, Tarakanov LA, ManeginYu.A. Molybdenum alloys. Moscow: Metallurgia (1975), in Russian.

[65] Evans AG, Zok FW.The physics and mechanics of fibre-reinforced brittle matrix composites.J Mater.Sci. 1994; 29: 3857 – 3896. DOI: 10.1007/BF00355946

[66] Zok FW.Developments in Oxide Fiber Composites.J. Am. Ceram. Soc.2006; 89:3309–3324. DOI: 10.1111/j.1551-2916.2006.01342.x

[67] M. K. Cinibulk, Hexaluminates as a cleavable fiber-matrix interphase: synthesis, texture development, and phase compatibility.Journal of the European Ceramic Society. 2000; 20:569-582.DOI:10.1016/S0955-2219(99)00255-1

[68] PopovaNA, TolstunAN. Formationofmonaziteinterphaseinoxide/oxidecompositeswithICM-fibres.SciencetoIndustry. 2007;#2:58-61 (in Russian).

[69] Mileiko ST, Kolchin AA, Novokhatskaya NI, Kiiko VM, Novikov IV.Quasi-plastic behaviour of brittle-matrix composites with oxide composite fibres. In: Proceedings of 16th European Conference on Composite Materials. Seville: July 2014.

[70] Mileiko ST, Novokhatskaya NI, Stolin AM, Bazhin PM. Producing silicide/molybdenum composites by using internal crystallization method. Composites and Nanostructures.2014; 6:185-197 (in Russian).

Chapter 5

Cocrystallization in Nutraceuticals

Renu Chadha, Yashika Bhalla,
Maninder Karan Vashisht and Kunal Chadha

Additional information is available at the end of the chapter

http://dx.doi.org/10.5772/59365

1. Introduction

Food materials that prevent and treat diseases are usually named as functional foods, designer foods, pharma foods and nutraceuticals [1]. Food has been used since ages as a source for the treatment and prevention of various disorders. Enrichment of table salt with iodine and wheat flour with iron/folic acid has been used with an intention to prevent iodine deficiency goiter and anemia for long. Similarly, food fortified with vitamin A has been found to be a feasible and cost-effective approach to reduce vitamin A deficiency [2].

Nutraceuticals very often named as " *functional foods* " provides the body with the essential amount of vitamins, fats, proteins and carbohydrates necessary for healthy survival. When functional food assists in the prevention and/or treatment of disease (s) other than deficiency conditions like anemia it is called a " nutraceutical " [3].

Nutritional therapy is a healing system using nutraceuticals or phytonutrients for the improvement of human health. It is well known that ancient civilizations used herbal products as a remedy for many diseases. However, specific term for a formulation describing food or its active ingredients, which have medicinal values, was eagerly required. Then, in 1989 term " Nutraceutical " was coined by merging the terms " Nutrition " and " Pharmaceutical " by Dr Stephen DeFelice [4]. " Nutraceutical " is a term used for nutritional supplement that is in the market with the objective to treat or prevent disease [5].Therefore, a "nutraceutical" is any matter that may be considered a food or part of a food and provides medical or health benefits. These food products may be categorized as isolated nutrients, dietary supplements and diets to genetically engineered "designer " foods, herbal products and processed foods such as cereals, soups and beverages [6]. Dietary supplements play a vital role in premature chronic disease appearance, disease progression, morbidity and mortality. Approximately 40-50% proportion in cardiovascular disorders, 35-50% proportion in cancers, and 20% proportion in osteoporosis is attributable to dietary factors [7]. These phytonutrients have a big market currently and annual growth rates of nutraceuticals are predicted to increase. This rate expected to be up to 25% for some products [8].

Nutraceuticals include many food and food products, including vitamins, soy products, glucosamine, chondroitin, and many polyphenols and flavonoids (resveratrol, ellagic acid, and quercetin). The vast majority of nutraceuticals are extracted from plant origins; such as fruits, vegetables, roots, and rhizomes. However, many nutraceuticals are derived from animal origins; vitamins and amino acids [9].

However these nutraceuticals suffer from poor solubility and subsequently bioavailability which hinders their formulation as well as utility. The cocrystallization of these nutraceutical is one of the novel strategies to overcome these problems. Cocrystallization of active pharmaceutical ingredients (APIs) with cocrystal formers (or coformers) has gained significant interest in drug-development since the resulting new solid forms are characterized by different physicochemical properties compared to the original API. The presence of many functional groups in APIs, which may form strong supramolecular synthons such as acid...acid, acid...pyridine, acid...amide, amide...amide, amide...pyridine N-oxide, O-H...O, O-H...N, N-H...O and N-H...N offers a great opportunity to design pharmaceutical cocrystals [10-15].Certainly, cocrystallization increases the diversity of solid-state forms of an API and enhances its pharmaceutical properties such as chemical stability, moisture uptake, mechanical behavior, solubility, dissolution rate and bioavailability. Hence, cocrystallization can directly impact scientific and legal aspects of drug development by providing alternative solid dosage forms and extended patent life.

Significant development on discovery of nutraceutical cocrystals has encouraged numerous studies on cocrystal formation and evaluation. This chapter presents the nutraceuticals cocrystals along with their preparation and characterization and crucial role in enhancing their pharmaceutical parameters.

2. Nutraceuticals: Role in therapeutics

Nutraceutical is the combination of " nutrition " and " pharmaceutical ". In order to modify and maintain normal physiological function nutraceuticals play an important role. The current population and the health trends are the main reasons for the growth and development in the field of nutraceuticals. These nutraceuticals fights against the major health issues of the century such as obesity, cardiovascular diseases, cancer, osteoporosis, arthritis, diabetes, cholesterol etc. In a nutshell, ' nutraceutical ' has lead to the new period of medicine and health, in which the food industry has become a research oriented sector [9].

Consumers being frustrated with the current status of disease-treatment approach with the modern medicines are seeking alternative safer and beneficial products and the bureaucracy of managed care makes nutraceuticals particularly appealing. " Let food be thy medicine and medicine be thy food ", quoted by Hippocrates is definitely the precept of today. Nutraceuticals are the promising class of natural products that demarcates between food and drugs to fade [8]. Even though the use of nutraceuticals by people has a long history, lately scientifically supported nutritional and medical evidence has permitted nutraceuticals to emerge as being potentially effective [16].

Research is now focused on the examination of nutritional food for their protective and disease preventing potential [17]. Nowadays research and development area are exploring different methods for standardization of the nutraceutical compounds or products, carefully develop and execute clinical studies to provide the health claims and benefits to the consumers as well as on the nutraceutical companies. Nutraceuticals have proven to be therapeutically active in the areas starting from common cold, weight problems, sleeping disorders, digestion, blood pressure, cholesterol control, pain killers, depression, anti-arthritic, cardio vascular diseases, diabetes ending up with prevention of certain cancers [18-21].

The nutraceuticals which include Polyunsaturated fatty acid (PUFA), Dietary fibres, Polyphenols, Spices, Prebiotics, Probiotics and Vitamins are the natural food sources (Figure 1) [22,23]. However, these nutraceuticals have high hydrophobicity and are sensitive to external agents such as air, light and oxidative enzymes which constitute a serious problem for their bioavailability.This further hamper their formulation and use. At this point their cocrystal can offer an opportunity to overcome all the issues of nutraceuticals and help in the exponential growth of nutraceuticals market.Cocrystal of nutraceuticals alone or in combination with other preventive and/or therapeutic strategies might become effective future drugs against the most common degenerative diseases. Cocrystals are engineered solids based on the concept of different forms having desired properties using hydrogen bonds,pi-pi stacking and van der waals interactions [9]

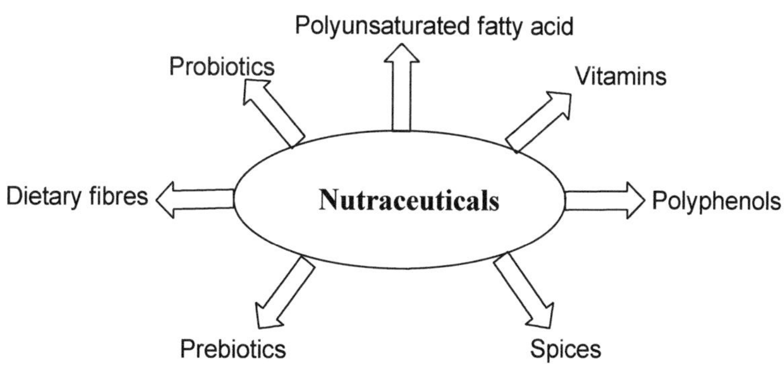

Figure 1. Types of Nutraceuticals

3. Cocrystallization

Cocrystals are based on crystal engineering approach which involves the concept of forming new solids (polymorph/solvates and cocrystals/salts) involving non covalent interactions.

Thus nutraceutical cocrystals are multi-component solid-state assemblies formed between a nutraceutical and a cocrystal former of GRAS status bound together in the crystal lattice by any type or combination of non-covalent intermolecular interactions.

Cocrystallization seeks to logically design new form of the nutraceuticals with preferred properties.Depending upon the various potential groups present in these nutraceutical molecules, following heterosynthons are possible (Figure 2) [24] :-

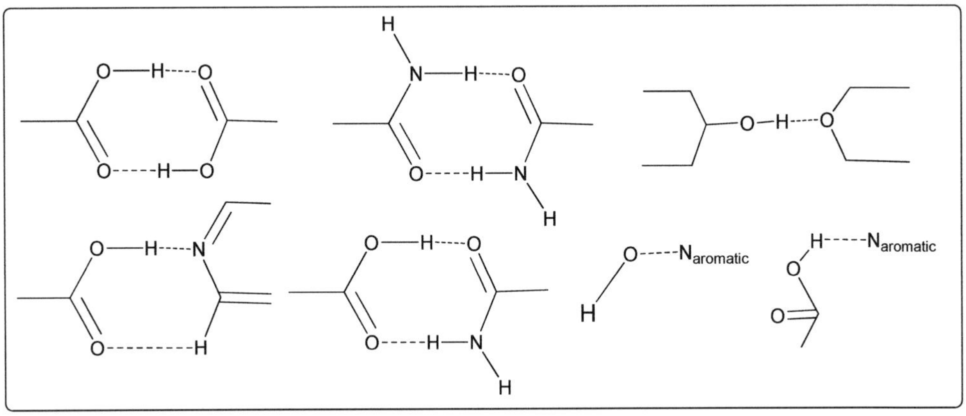

Figure 2. Typical Hydrogen Bonds Utilized in Cocrystals

Cocrystallization is at present achieved by a number of techniques including solvent evaporation, dry and wet (solvent drop addition) solid-state grinding, cooling, growth from melts, and slurry conversion. This section covers the various methods of preparation and characterization of the cocrystals.

Different techniques for the preparation of co crystals (Figure 3) [25] are described below:-

1. Solvent evaporation technique :- This is the most commonly used technique in the field of cocrystallization, in this technique stoichiometric amount of the conformer and the drug are dissolved in a common solvent. It is based on the principle that, when different molecules of complementary functional groups afford hydrogen bonds that are more favorable than each of the individual molecular components, the product formed is likely to be thermodynamically favored [26].

2. Solid-state grinding technique :- This technique commonly known as mechanical milling or neat grinding technique.in this method the two cocrystal formers are taken in stoichiometric amounts and ground together using a mortar and pestle, using a ball mill, or using a vibratory mill. Normal grinding time ranges from 30-60 min. It has been reported that cocrystal was first produced by one technique which may be used as seeds to obtain that cocrystal by another method, thus possibly facilitating XRD structure determination via single-crystal growth. In another case, cocrystal structure determination was achieved by preparing only as crystalline powder by grinding [27]

http://dx.doi.org/10.5772/59365

Although co-crystal formation by solid-state grinding has been known for some time and a late 19th century report is often cited as the earliest reference to such a procedure, the recent technique of adding small amounts of solvent during the grinding process has been shown to enhance the kinetics and facilitate co-crystal formation and as lead to increased interest of solid-state grinding as a method for co-crystal preparation [28].

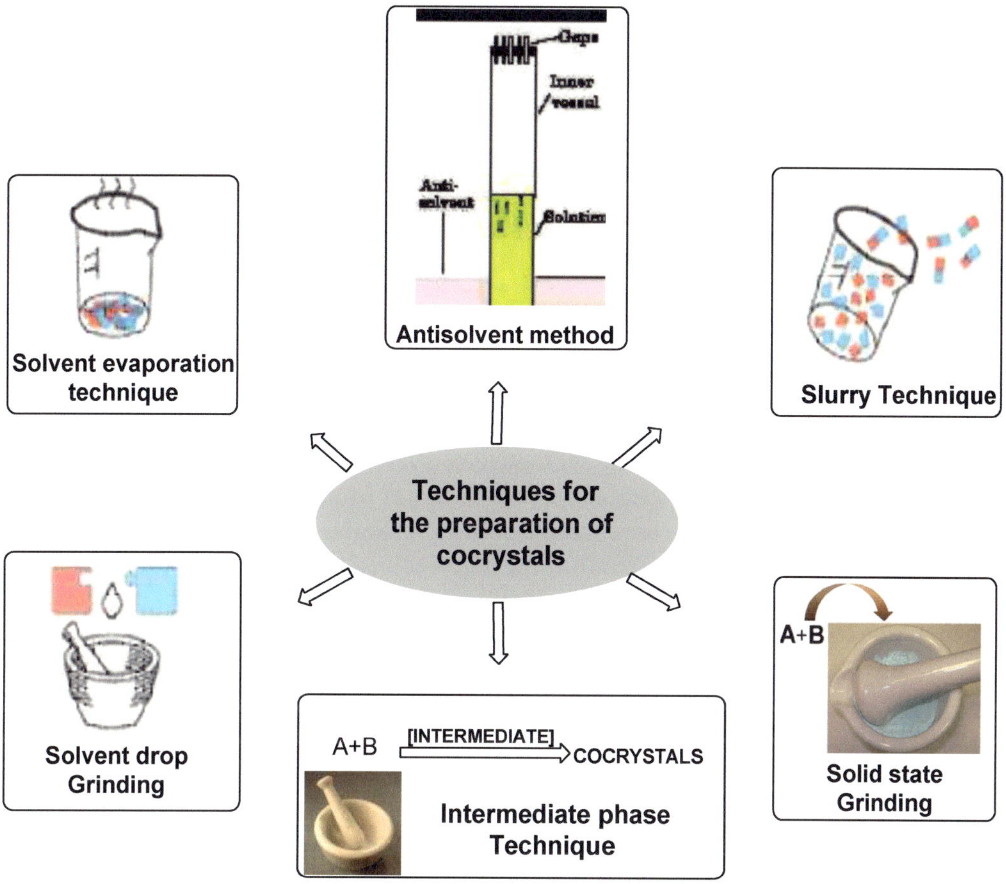

Figure 3. Techniques for the Preparation of Cocrystals

3. Slurrying technique:-Slurries-induced formation of cocrystalline phase is among two or more active solid materials or between the active solid materials and the excipients. Equimolar proportion of the two coformers are dissolved in small amount of different solvents at ambient temperature and allowed to stir for some days. The resulting solution is slowly evaporated at room temperature during 48 hours to promote cocrystallization [29]..Further the solvent is allowed to decant and the solid material is dried under a flow of nitrogen for 5 min. The remaining solids can be characterized using powder X-ray diffraction (PXRD) [30].

4. Solvent drop technique:-This technique also known as liquid assisted grinding or kneading, involving the grinding of stoichiometric amounts of coformers assisted with a small amount of liquid. This method was developed in order to enhance the rate of cocrystal formation, It also has advantages over solid state grinding such as increased yield, ability to control polymorph production, better product crystallinity, and applies to a significantly larger scope of cocrystal formers [31].

5. Supercritical fluid technology :-Supercritical fluids is another method by which cocrystals are synthesized.Supercritical fluid technology allows a single-step generation of particles that are difficult or even impossible to obtain by conventional techniques.The unique properties of different super critical fluids assist in generation of pure and dried cocrystals. [32]

6. By using intermediate phase:-Cocrystals can be formed also by using intermediate phases. In the course of time the use of a hydrate or an amorphous phase as an intermediate during synthesis in a solid-state route has proven successful in forming a cocrystal. Besides this, the use of a metastable polymorphic form of one cocrystal former can be used. The metastable form in this technique acts as an unstable intermediate on the nucleation pathway to a cocrystal.

7. Antisolvent addition:-Antisolvent addition also known as vapor diffusion is the methods in which antisolvent is used to synthesize cocrystals.This is one of the methods for precipitation or recrystalization of the two co-crystal former. Solvents consist of buffers (pH) and organic solvents. For example, in the preparation of co-crystals of aceclofenac using chitosan, the chitosan solution was prepared by soaking chitosan in glacial acetic acid. A stoichiometric amount of the drug was suspended in chitosan solution by using high dispersion homogenizer. Further, the dispersion was added to distilled water or sodium citrate solution to precipitate chitosan on drug [33].

3.1. Characterization of cocrystals

This of great importance which can be done using different analytical techniques.Characterization of cocrystals starts from melting point determination and might end up to the technique (Figure 4) employed for the determination of cocrystals.

Thermal analysis:- Differential Scanning Calorimetry (DSC) plays a major role in determining the thermal property testing of cocrystals.DSC is the technique used to observe enthalpy of melting (pharmaceutical cocrystal an overview)during various crystallization events.

Besides this, DSC has recently been used as a screening tool for rapid cocrystal screening [34, 35].It has also been reported that when theophylline and nicotinamide were cogrinded, and from the DSC results it observed that an endothermal peak at about 126.4°C decreased and finally disappeared. A melting endotherm at 171.6°C, which differs from the melting points of either theophylline (271.4°C) or nicotinamide (128.2° C), which indicates that as the cogrinding of theophylline and nicotinamide proceeded, a new phase was formed between theophylline and nicotinamide during the solid- state grinding [36].

http://dx.doi.org/10.5772/59365

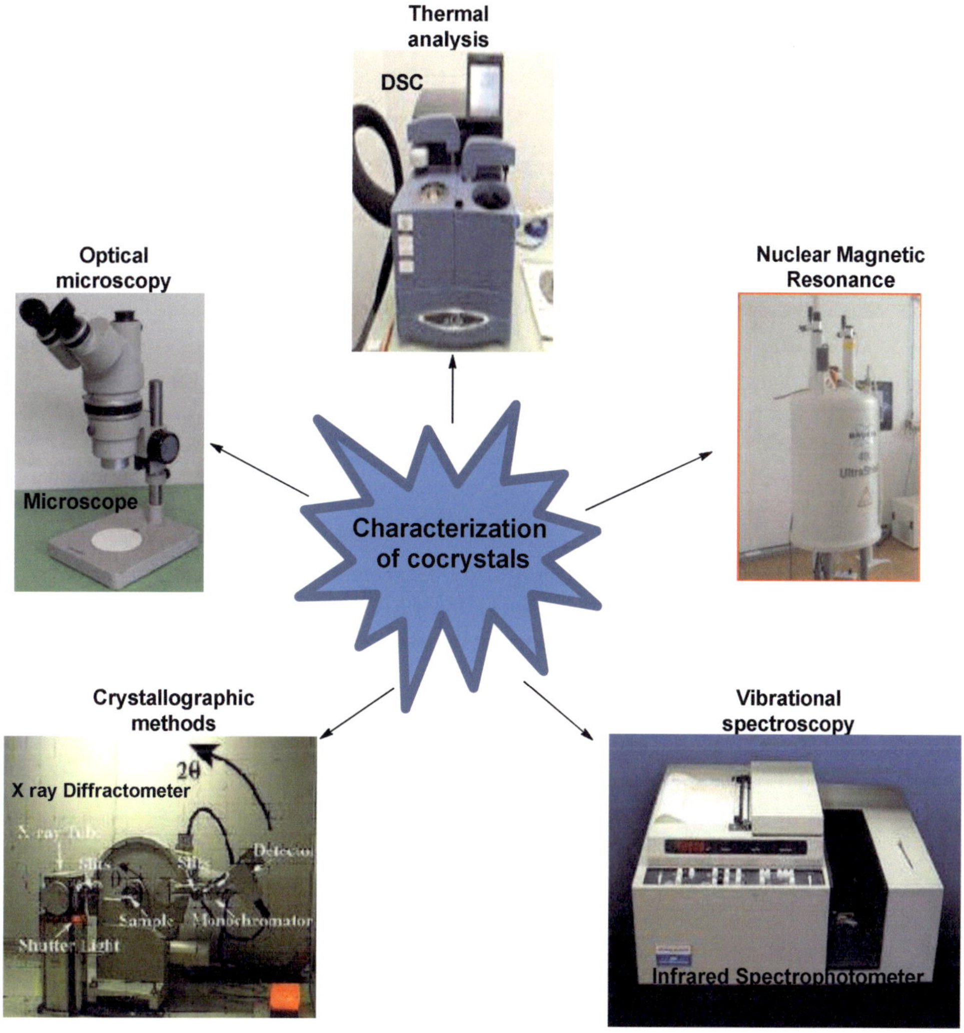

Figure 4. Characterization of Cocrystals

Vibrational spectroscopy:-For the characterization of cocrystals molecular motions by use of vibrational spectroscopy are observed. This method includes infrared absorption spectroscopy.Significant changes in the vibrational frequency are observed for the cocrystals and individual constituents confirming the formation of cocrytals. It can be a very useful technique in differentiaiting cocrystals from salts when a carboxylic acid is involved in hydrogen bonding [37].

It has been reported that in the case of indomethacin and saccharin cocrystals, the O–H stretching of the carboxylic acid group in indomethacin and the N–H stretching of cyclic imide group in saccharin appeared at 3,350.5 and 3,129.1 cm in Indomethacin and saccharin cocrystals, implied the formation of new phase [38].

Crystallographic methods: - Includes single X- ray diffraction (SXRD) as well as PXRD. The single X-ray diffraction study offers atomic positions and complete structural information. Powder X-ray diffraction studies using microcrystalline samples become a necessary tool to verify the formation of cocrystal in cases where obtaining single crystal suitable for study becomes bottleneck.SXRD is commonly used as supporting information about the formation of a new crystalline phase as in the case of Ciprofloxacin and Norfloxacin cocrystallization [39].

Optical microscopy:-Optical microscopy is the technique used to study crystal morphology. This feature is reflected in the outer appearance of crystals that can be observed by simple hand lens or microscope. To add on, a detailed study can be performed using polarizing optical microscopy, electron microscopy and thermal microscopy.Scanning electron microscopy (SEM), is also used to characterize the surface morphology of the particles. SEM images a sample by scanning it with a high-energy beam of electrons in a raster scan pattern. It is applied to determine the cocrystal micrograph and determine the particle morphology in many examples [40-43].

Solid-state nuclear magnetic resonance - These days solid-state NMR is also used for characterization. It studies the chemical environment of the nuclei which is different in cocrystals because of magnetic non-equivalence. In different cocrystals, resonance peaks for the magnetically non-equivalent nuclei will differ. Dipolar correlation experiments between spin pairs such as ^{1}H-^{1}H,^{1}H-^{13}C, and ^{19}F-^{13}C are useful to study hydrogen bonding, intermolecular contacts, and spin diffusion to connect individual molecules together in a crystal structure and rapidly establish molecular association [44].

3.2. Case studies of some nutraceutical cocrystals

Nutraceuticals such as flavonoids and vitamins have been investigated as candidates for cocrystallization studies to improve their pharmaceutical properties [45].

Flavonoids are natural products and are often studied because of their potent antioxidant and free radical scavenging activities. Nutraceutical cocrystals are associated with higher *in vitro* dissolution rate and *in vivo* bioavailability of their other solid forms, as reported by Smith et al. for quercetin [46] and Epigallocatechin gallate (EGCG). [47].

It has been reported that cocrystallization of flavonoids with 1,4-diazobicyclo [2.2.2]octane (DABCO) resulted in increase in the complexity of these molecules as the number of substituents on the flavonoid backbone increased. Various preparation and properties of flavonoid interactions to the formation of cocrystals with APIs were reported [48].

3.2.1. Protocatechuic acid

Protocatechuic acid (3,4-dihydroxybenzoic acid, Figure 5) is a phenolic acid and belongs to class of polyphenols (Nutraceutical). Protocatechuic acid imparts various pharmacological activity and these effects are due to their antioxidant activities, along with other possible mechanisms, such as anti-inflammatory properties and interaction with several enzymes.

http://dx.doi.org/10.5772/59365

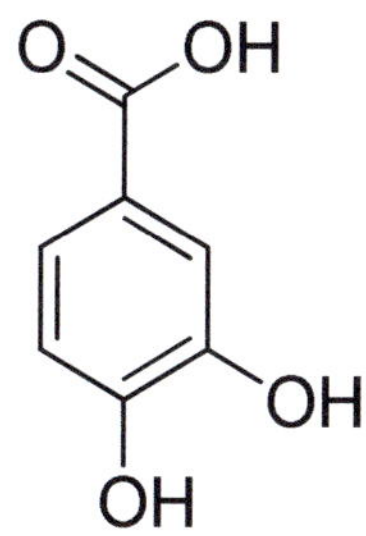

Figure 5. Protocatechuic Acid

S.NO.	Co-former	PA* + Co-Former
1	Caprolactam	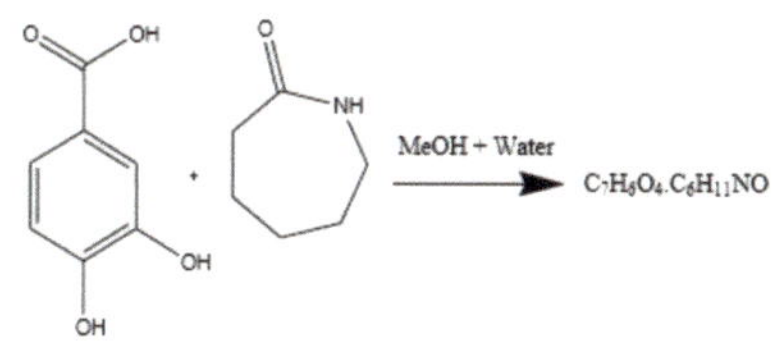 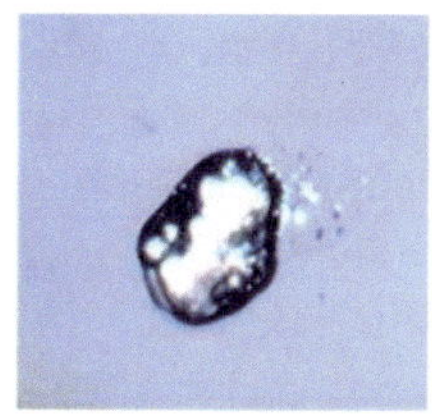
2	Isonicotinamide	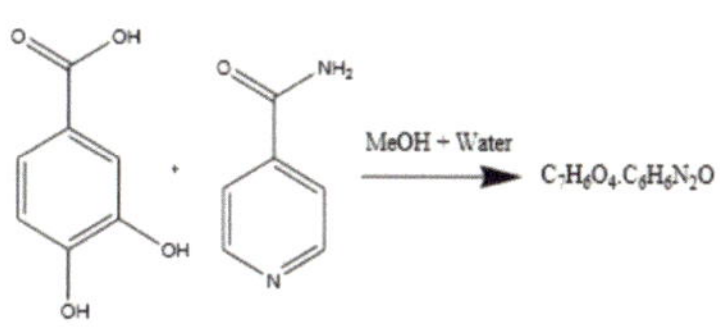 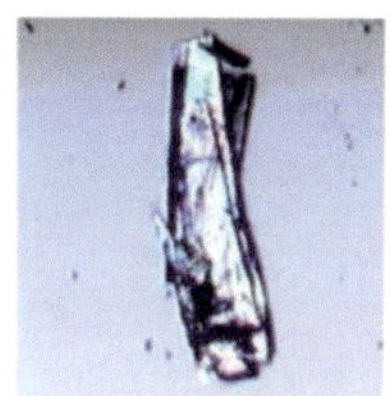
3	Isonicotinic Acid	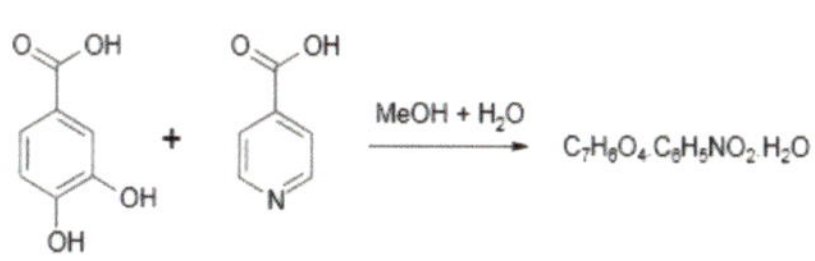 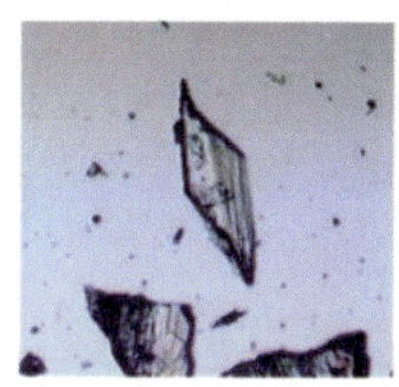
4	Theophylline	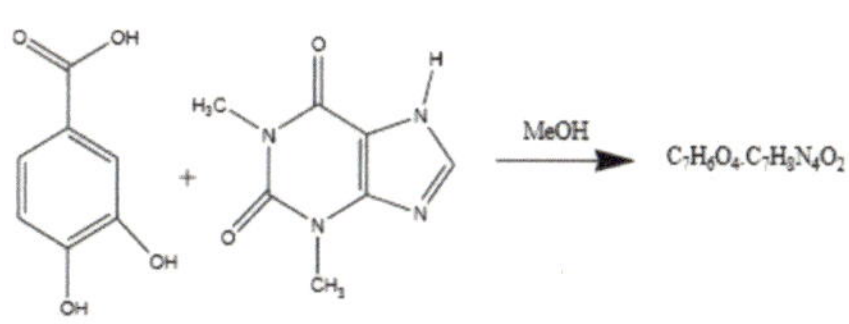 

Table 1. Single crystals of Protocatechuric acid with different coformers [49]

Further in this context, novel 1:1 cocrystals of protocatechuic acid (strong antioxidant) with pharmaceutically accepted molecules (cocrystal formers) such as caprolactam, isonicotinamide, isonicotinic acid, theophylline, nicotinamide and theobromine (Table No. 1) have been obtained by slow evaporation of stoichiometric amounts of starting materials in an appropriate solvent and they were removed from their mother liquors before complete evaporation of the solvent. Cocrystallization via grinding and slurry conversion was also successful to produce 1:1 cocrystals of protocatechuic acid with caprolactam, isonicotinamide, isonicotinic acid, theophylline, nicotinamide and theobromine. The resultant cocrystals were characterized by FTIR, DSC, PXRD, single crystal X-ray diffraction and TGA. [49]

3.2.2. *Quercetin*

Quercetin (3,3',4',5-7-pentahydroxyflavone) (Figure 6), chemically similar to the glycoside rutin, is a unique flavonoid that has been extensively studied by researchers. It has shown to be therapeutically useful in allergic conditions, including asthma and hay fever, eczema, and hives. The frequent link is its ability to mediate production and manufacture of pro-inflammatory compounds.Cocrystal of Quercetin, isonicotinic acid and water (1:1:1) and cocrystal of quercetin, theobromine and water (1:1:2) were obtained (Table no. 2). Quercetin theobromine dihydrate cocrystal resulted in 1.5 fold increase in solubility of quercetin. Quercetin-caffeine-methanol cocrystal was observed after 30 days of slow evaporation of the Quercetin-caffeine-methanol cocrystal solvate.The resulting crystal structure of single crystal of quercetin-caffeine-methanol cocrystal solvate showed that the imide group and the aromatic nitrogen of caffeine interact with the hydroxyl groups of quercetin [50].

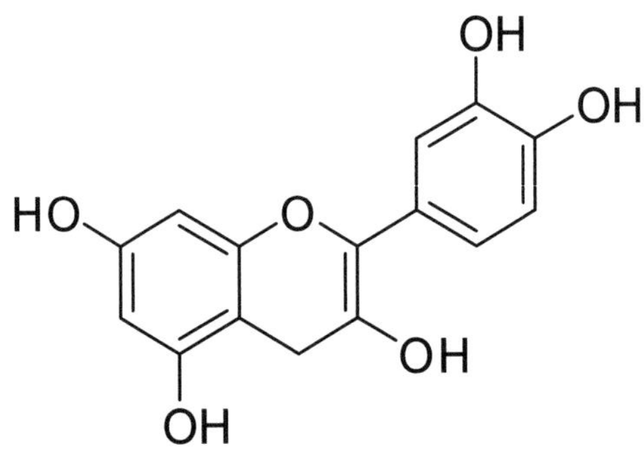

Figure 6. Quercetin

In another attempt for the synthesis of cocrystal of Quercetin with isonicotinamide, quercetin dehydrate and isonicotinamide are dissolved in methanol (5 ml) and heated until a clear solution was obtained. Slow evaporation of this solution in refrigerator resulted in 1:1 crystals after 2 days. Isonicotinamide molecules interact with quercetin molecules via, CO...OH and NH...OH supramolecular heterosynthons and OH...OH supramolecular homosynthons. [50]

Four cocrystals of quercetin (QUE): quercetin: caffeine (QUECAF), quercetin: caffeine: methanol (QUECAF· MeOH), quercetin: Isonicotinamide (QUEINM) and quercetin: theobro-

S.NO.	Co-Former	Quercetin + co-Former
1	Isonicotinic acid + water	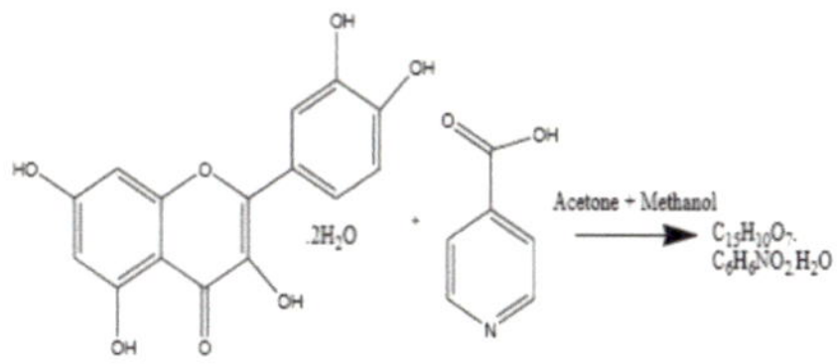 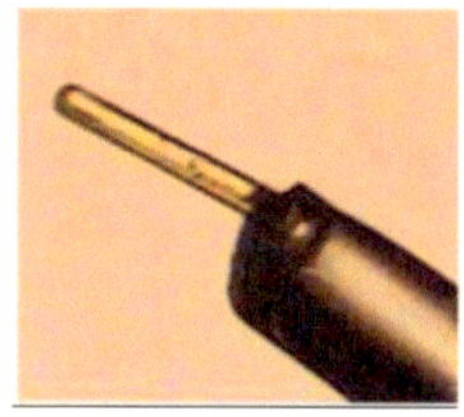
2	Theobromine + water	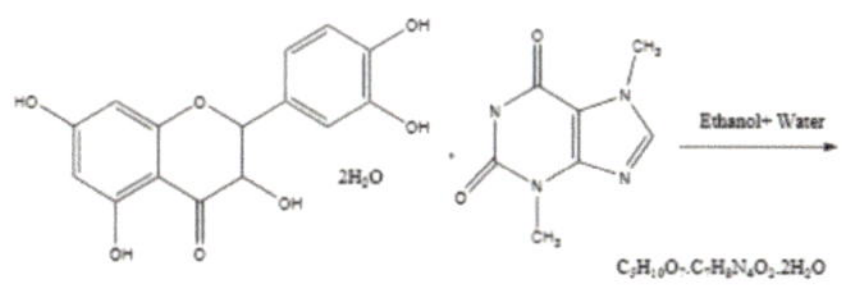 

Table 2. Co- crystals of Quercetin with different coformers [50]

mine dihydrate (QUETBR·2H2O) were prepared by slow evaporation (solution crystallization) method and each of these cocrystals exhibited pharmacokinetic properties that are superior to those of quercetin alone. The QUECAF and QUECAF·MeOH cocrystals increased the solubility of QUE by 14-and 8-fold when compared to QUE dihydrate. Further, the cocrystals outperformed QUE dihydrate with increases in bioavailability up to nearly 10-fold [51].

3.2.3. Hesperetin

Hesperetin (Figure 7) which is the aglycone part of hesperedin and is therapeutically very potential molecule as it shows antioxidant, antiallergic, antimutagenic and *in vitro* studies anti cancer activity. Cocrystallization of hesperetin with isonicotinamide resulted in a 1:1 cocrystal. The supramolecular synthon formed in the cocrystal includes OH---N hydrogen bond between the nitrogen atom of isonicotinamide and the OHa of the adjacent hesperetin molecule. Crystallization of hesperetin with nicotinic acid results in two 1:1 cocrystals in which the nicotinic acid exists as a zwitterionic state. Cocrystallization has lead to the generation novel cocrystals of hesperetin with pharmaceutically acceptable molecules such as isonicotinamide and nicotinic acid [52].

3.2.4. Pterostilbene

Pterostilbene (Figure 8) is also a nutraceutical which is found in nature in a number of tree barks and a variety of berries, including grapes, as well as plants commonly used in traditional folk medicine. Cocrystals of pterostilbene comprises of coformers such as: caffeine, carbamazepine. Three cocrystals of a 1:1 stoichiometric molar ratio of pterostilbene with caffeine and carbamazepine were obtained and characterized by crystallographic (XRPD, single-crystal) and thermoanalytical (TGA, DSC) techniques. Physical stability of the reported cocrystals with respect to relative humidity was established to be significantly improved in relationship to

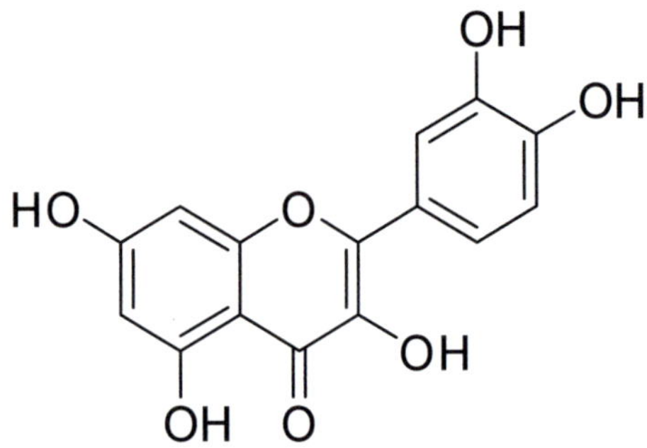

Figure 7. Hesperetin

S.NO.	Co-Former	Single crystal of Hesperetin +conformer
1	Isonicotinamide	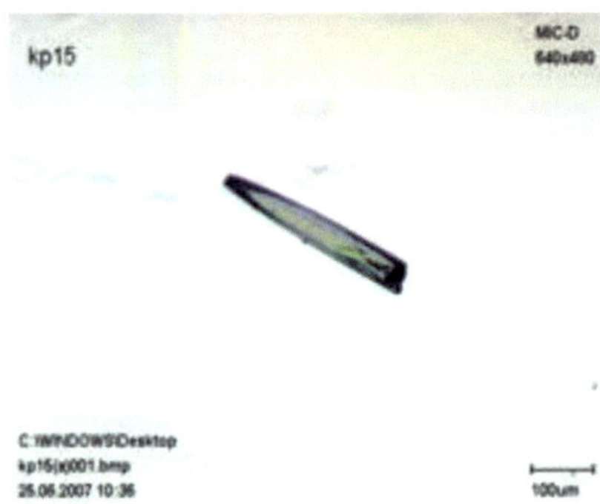
2	Nicotinic acid.	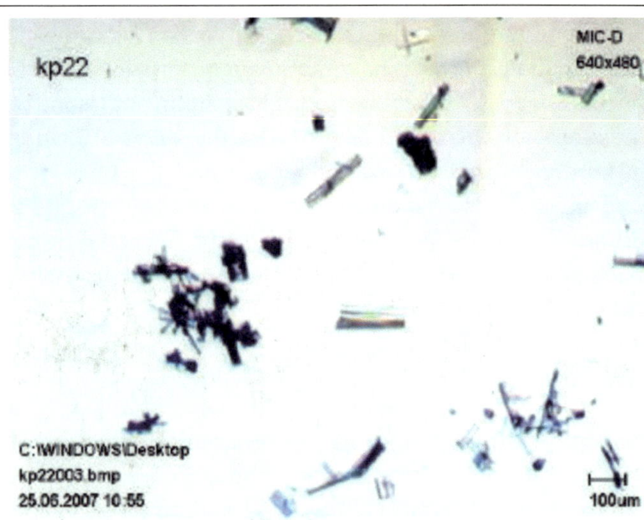

Table 3. C ocrystalsof Hesperetin with different coformers [52]

caffeine or carbamazepine. The carbamazepine: pterostilbene cocrystal was found to be stable upon slurrying in water for 3 days and its solubility was 7× lower than carbamazepine dihydrate and 2.5× lower than pterostilbene [53].

Another study on pterostilbene cocrystallization with pharmaceutically acceptable coformers piperazine and glutaric acid was reported by Bethune et al. [54]. Cocrystals of a 2:1 and 1:1 stoichiometric molar ratio of pterostilbene with piperazine or glutaric acid were synthesized and evaluated for physical stability with respect to humidity and temperature as well as kinetic solubility.The pterostilbene/piperazine cocrystal disclose a 6-fold increase in pterostilbene concentration compared with the solubility of pterostilbene.

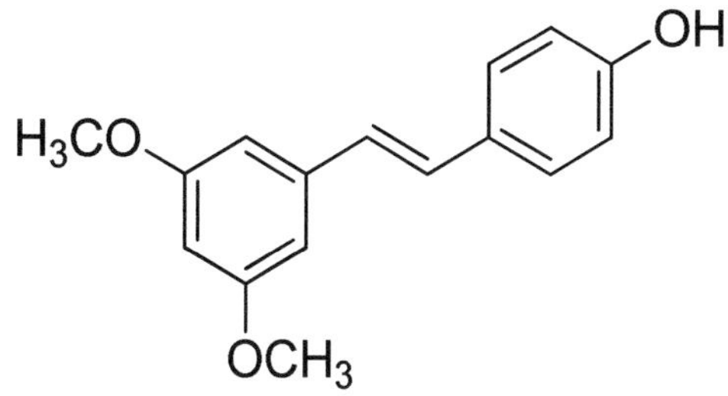

Figure 8. Pterostilbene

3.2.5. p-Coumaric

p-Coumaric (Figure 9) acid is a phytochemical and nutraceutical compound.Cocrystals synthesized with *p*-coumaric acid are explored with xanthine compounds, caffeine and theophylline. Four cocrystals of *p*-Coumaric acid with caffeine (1: 1 and 1: 2 stoichiometric ratios) and was synthesized and their structures determined by single-crystal X-ray crystallography. The two theophylline cocrystals display synthon polymorphism, where both structures possess a carboxylic acid–imidazole heteromeric synthon [55].

In another work reported for polymorphic cocrystals of p-Coumaric acid, the nutraceutical was cocrystallized with nicotinamide, producing three polymorphic 1:1 cocrystals and one 2:1 (p-Coumaric acid-nicotinamide) cocrystal. Conversion of the 1:1 cocrystals to the 2:1 cocrystal in water was revealed by *in situ* dispersive Raman spectroscopy. An enantiotropic relationship is believed to be existing between two of the 1:1 cocrystals; the third polymorph was likely to be metastable [56].

3.2.6. Curcumin

Curcumin, the most important curcuminoid extracted from Curcuma *longa* rhizomes, was commonly used as a spice until its medicinal properties in Indian and Chinese systems of medicine were known and documented.Curcumin (Figure 6) is extensively used as a drug but its utility is hindered due to poor aqueous solubility.

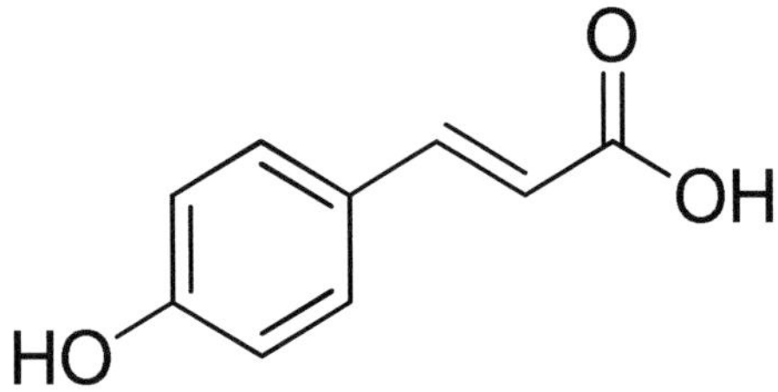

Figure 9. p-Coumaric acid

It has been reported that the cocrystals of curcumin with resorcinol and pyrogallol were obtained by solvent drop method. Cocrystals of Curcumin–resorcinol and curcumin–pyrogallol were obtained in stoichiometric ratio of 1: 1and were characterized by X-ray diffraction, thermal analysis, FT-IR, FT-Raman, and solid-state 13C NMR spectroscopy. The melting point of the cocrystals was found to be between that of curcumin and the conformer,it has been observed that the lower melting cocrystal was more soluble than higher melting.The dissolution studies revealed that the dissolution rates of curcumin–resorcinol and curcumin–pyrogallol in 40% EtOH–water are ~5 and ~12 times faster than that for curcumin [57].

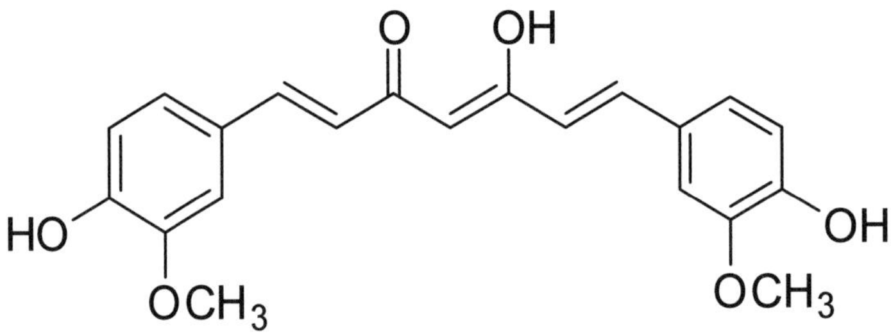

Figure 10. Curcumin

3.2.7. Citric acid

Citric acid (2-hydroxy-1, 2, 3-propanetricar-boxylic acid) is a weak tricarboxylic acid that is naturally concentrated in citrus fruits. Citric acid is commonly used as a food additive to provide acidity and sour taste to foods and beverages. Citrate salts of various metals are used to deliver minerals in biologically-available forms; examples include dietary supplements and medications. Citric acid is known as a highly soluble and safe nutraceutical. Since citric acid is highly soluble it may be used as conformer in the field of cocrystallization.

It was reported that Citric acid (Figure 11) and paracetamol in 2:1 stoitiometric ratio formed cocrystals. Notable changes were observed in the IR spectra of the cocrystal. IR spectra of cocrystal were different from that of the raw materials by several vibrational frequencies. The asymmetric unit of the crystal has two paracetamol molecules hydrogen-bonded to the citric acid; one of these behaves as a phenolic-OH hydrogen bond donor to the carbonyl of a carboxylic acid arm of citric acid. In contrast, the other phenolic-OH acts as a hydrogen bond acceptor from the quaternary C–OH of citric acid [58].

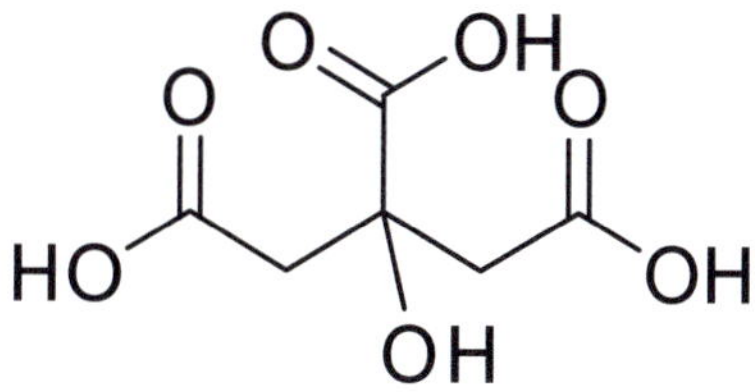

Figure 11. Citric acid

Dalia et al., had reported about the formation of a novel cocrystals of citric acid resulting in the formation of single crystals (Table no. 4). Cocrystal of citric acid in 2:1 ratio with isonicotinamide was reported and X-ray crystal structure indicates that molecules of *iso*-nicotinamide form hydrogen bonds with molecules of citric acid [59].

S.NO.	Co-Former	Single crystal
1.	Isonicotinamide	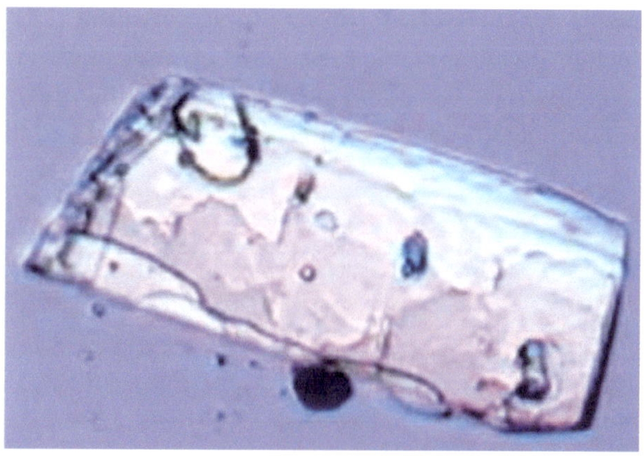

Table 4. Cocrystal of Citric acid

3.2.8. Gossypol

Gossypol (Figure 12) is a natural product occurring as biphenolic compound derived from the cotton plant (genus *Gossypium*). Wang et al. in 2009 reported the (-)-gossypol cocrystals with a C1–8 carboxylic acid or C1–8 sulfonic acid which are inhibitors of anti-apoptotic Bcl-2 family proteins. The invention also relates to inducing apoptosis in cells and for sensitizing cells to the induction of apoptotic cell death by the cocrystals of (-)-gossypol with a C1–8 carboxylic acid or C1–8 sulfonic acid [60].

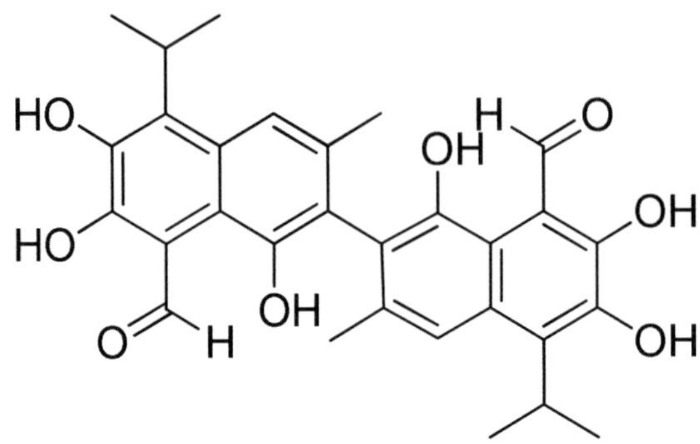

Figure 12. Gossypol

3.2.9. Fisetin, Luteolin, Genistein

Fisetin, Luteolin and Genistein (Figure 13) are natural polyphenolic compounds belonging to nutraceuticals.Crystallization studies were carried on these molecules with nicotinamide and isonicotinamide as coformers. Five cocrystals were isolated, characterized by X-ray single-crystal diffraction, FT-Raman spectroscopy, thermal analysis (DSC and TG–DTA), ^{1}H NMR in solution and compared in terms of supramolecular motifs. Fisetin–nicotinamide (1: 2) ethanol hemisolvate (**FisNam**), fisetin–isonicotinamide (1:1) (**FisInam**), two polymorphic forms of luteolin–isonicotinamide (1: 1) (**LutInam**, **LutInam2**) and genistein–nicotinamide (1:1) monohydrate (**GenNam**) cocrystals reveal the presence of an OH····N (atom) heterosynthon between an O7 hydroxyl moiety of a flavonoid and the pyridyl ring of a conformer [61].

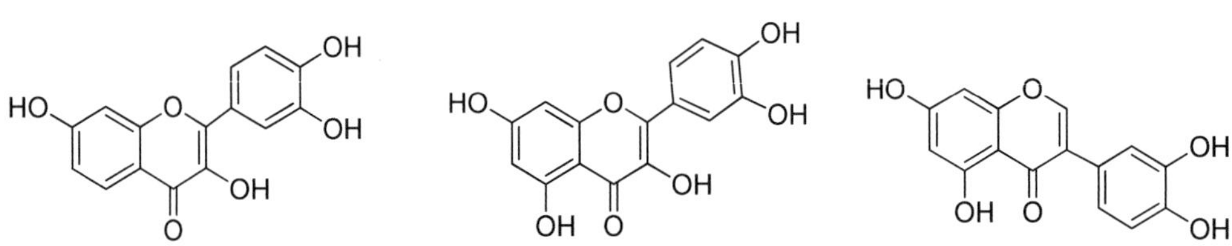

Figure 13. Fistelin, Luteolin and Genistein

Genistein (Figure 13) is a phytoestrogen and belongs to the category of isoflavones. Genistein and Caffeine formed 1:1 cocrystalline phase, which was formed by means of a solvent-drop grinding experiment and isolated later in a solution-evaporation approach. Resulting cocrystal was characterized by X-ray single-crystal and powder diffraction and the cocrystals were explored in terms of thermal stability and Hirshfeld surfaces. A scale-up procedure by slurry method, facilitated solubility studies. Neutral forms of both compounds cocrystallize in a common P21/c space group of the monoclinic crystal system. Crystal lattice discloses formation of molecular layers, formed by O–H_ _ _O, O–H_ _ _N and C–H_ _ _O-type contacts between genistein and caffeine molecules, while stabilization of the three-dimensional crystal lattice is offered by pi - pi interactions. Dissolution studies showed maximum solubility of the cocrystalline phase in a 50:50 v/v ethanol–water medium, attained 0.861 mg/mL after 8 h, revealing some degree of enhancement as compared to parent genistein, maximum solubility of which was also reached after 8 h and equaled to 0.588 mg/mL [62].

4. Conclusion

These natural products are of interest because of their proposed health-promoting effects as antioxidants and anticarcinogens. In spite of the immense biological activities revealed by nutraceuticals still these phytogenes cannot be exploited due to their biopharmaceutical issues which hinder their utility. However, their high hydrophobicity and sensitivity to external agents such as air, light and oxidative enzymes constitute a serious problem for their bioavailability and formulation.. However, bioavailabilty and solubility troubles coupled with these natural entities can be improved by formulating them into cocrystals. The cocrystallization strategy may optimize some potential nutraceuticals. Cocrystal of nutraceuticals alone or in combination with other preventive and/or therapeutic strategies might become effective future drugs against the most common degenerative diseases. Though studies on co crystallization of nutraceutical molecules are going on, still there is a lot to be explored so as to utilize these potential molecules for severe disease.

Author details

Renu Chadha*, Yashika Bhalla, Maninder Karan Vashisht and Kunal Chadha

*Address all correspondence to: renukchadha@rediffmail.com

University Institute of Pharmaceutical Sciences, Panjab University, Chandigarh, India

References

[1] Kwak NS, Jukes DJ. Functional foods. Part 1: the development of a regulatory concept. Food Control. 2001;12 (2)99-107

[2] Bele AA, Khale A. An approach to a nutraceutical. Research Journal of Pharmacy and Technology. 2013; 6 (10)1161-1164

[3] Kalra EK. Nutraceutical-definition and introduction. American Association of Pharmaceutical Scientists. 2003; 5 (3) 27-28.

[4] Brower V. Nutraceuticals: poised for a healthy slice of the healthcare market? Nature biotechnology. 1998;16 728-732.

[5] Zeisel SH. Regulation of " nutraceuticals". Science 1999;285:1853-1855.

[6] Eskin M, Tamir S. Dictionary of nutraceuticals and functional foods: CRC Press; 2004.

[7] Gupta S, Yadav SK, Patil S. Nutraceutical: a bright scope and oppourtunity of Indian healthcare market. International Journal of Research and Development in Pharmacy and Life Sciences 2013: 2 (4) 478-481.

[8] Adelaja AO, Schilling BJ. Nutraceuticals: blurring the line between food and drugs in the twenty-first century. Choices. 1999;14 (4).

[9] Das L, Bhaumik E, Raychaudhuri U, Chakraborty R. Role of nutraceuticals in human health. Journal of food science and technology. 2012;49 (2) 173-183.

[10] Reddy LS, Babu NJ, Nangia A. Carboxamide–pyridine N-oxide heterosynthon for crystal engineering and pharmaceutical cocrystals. Chemical Communications. 2006 (13) 1369-1371.

[11] Bhogala BR, Basavoju S, Nangia A. Three-component carboxylic acid-bipyridine lattice inclusion host. Supramolecular synthesis of ternary cocrystals. Crystal Growth & Design. 2005;5 (5) 1683-1686.

[12] Reddy LS, Basavoju S, Vangala VR, Nangia A. Hydrogen bonding in crystal structures of N, N'-Bis (3-pyridyl) urea. Why is the NH tape synthon absent in diaryl ureas with electron-withdrawing groups? Crystal Growth & Design. 2006;6 (1) 161-173.

[13] Shattock TR, Vishweshwar P, Wang Z, Zaworotko MJ. 18-Fold interpenetration and concomitant polymorphism in the 2: 3 Co-crystal of trimesic acid and 1, 2-bis (4-pyridyl) ethane. Crystal Growth & Design. 2005; 5 (6) 2046-2049.

[14] Bhogala BR, Nangia A. Cocrystals of 1, 3, 5-Cyclohexanetricarboxylic Acid with 4, 4'-bipyridine homologues: acid pyridine hydrogen bonding in neutral and ionic complexes. Crystal Growth & Design. 2003;3 (4) 547-554

[15] Aakeröy CB, Hussain I, Desper J. 2-Acetaminopyridine: A highly effective cocrystallizing agent. Crystal Growth & Design. 2006;6 (2) 474-80.

[16] Dillard CJ, German JB. Phytochemicals: nutraceuticals and human health. Journal of the Science of Food and Agriculture. 2000;80 (12) 1744-1756

[17] Nicoli M, Anese M, Parpinel M. Influence of processing on the antioxidant properties of fruit and vegetables. Trends in Food Science & Technology. 1999;10 (3) 94-100.

[18] Pandey M, Verma RK, Saraf SA. Nutraceuticals: new era of medicine and health. Asian Journal Pharmaceutical Clinical Research. 2010;3 (1) 11-15.

[19] Sami Labs—pioneer in nutraceuticals August 05, 2002. The Hindu Newspaper :http://www.hinduonnet.com/thehindu/biz/2002/08/05/stories/2002080500040200.htm

[20] Espín JC, García-Conesa MT, Tomás-Barberán FA. Nutraceuticals: facts and fiction. Phytochemistry. 2007;68 (22) 2986-3008.

[21] Jang M, Cai L, Udeani GO, Slowing KV, Thomas CF, Beecher CW, et al. Cancer chemopreventive activity of resveratrol, a natural product derived from grapes. Science. 1997;275 (5297) 218-220.

[22] Kalia AN. Textbook of Industrial Pharmacognocy, CBS publisher and distributor, New Delhi, 2005, pp. 204–208

[23] Kokate CK, Purohit AP, Gokhale SB. Nutraceutical and Cosmaceutical. Pharmacognosy, 21st edition, Pune, India: Nirali Prakashan,2002 pp. 542–549

[24] Qiao N, Li M, Schlindwein W, Malek N, Davies A, Trappitt G. Pharmaceutical cocrystals: an overview. International Journal of Pharmaceutics. 2011;419 (1):1-11.

[25] Review on cocrystal as an approach with newer implications in pharmaceutical field

[26] Jayasankar A, Somwangthanaroj A, Shao ZJ, Rodríguez-Hornedo N. Cocrystal formation during cogrinding and storage is mediated by amorphous phase. Pharmaceutical Research. 2006;23 (10) 2381-2392.

[27] Zaini E, Sumirtapura YC, Soewandhi SN, Halim A, Uekusa H, Fujii K. Cocrystalline phase transformation of binary mixture of trimethoprim and sulfamethoxazole by slurry technique. Asian Journal of Pharmaceutical and Clinical Research. 2010;3 (4).

[28] Blagden N, De Matas M, Gavan P, York P. Crystal engineering of active pharmaceutical ingredients to improve solubility and dissolution rates. Advanced Drug Delivery Reviews. 2007;59 (7):617-630.

[29] Tiekink ERT, Vittal JJ (Editors). Frontiers in Crystal Engineering, John Wiley & Sons, Ltd, Chapter 2: Crystal Engineering of Pharmaceutical Co-crystals, 2006, pp. 25-50

[30] Yadav A, Shete A, Dabke A, Kulkarni P, Sakhare S. Co-crystals: a novel approach to modify physicochemical properties of active pharmaceutical ingredients. Indian Journal of Pharmaceutical Sciences. 2009;71 (4):359.

[31] Jung J, Perrat M. Particle design using supercritical fluid: literature and patent survey. Journal of Supercritical Fluids. 2001;20 179-219.

[32] Basavoju S, Boström D, Velaga SP. Indomethacin–saccharin cocrystal: design, synthesis and preliminary pharmaceutical characterization. Pharmaceutical Research. 2008;25 (3) 530-541.

[33] Mutalik S, Anju P, Manoj K, Usha AN. Enhancement of dissolution rate and bioavailability of aceclofenac: a chitosan-based solvent change approach. International Journal of Pharmaceutics. 2008;350 (1) 279-290.

[34] Lu E, Rodríguez-Hornedo N, Suryanarayanan R. A rapid thermal method for cocrystal screening. Crystal Engineering Communication. 2008;10 (6) 665-668.

[35] Mohammad MA, Alhalaweh A, Velaga SP. Hansen solubility parameter as a tool to predict cocrystal formation. International Journal of Pharmaceutics. 2011;407 (1) 63-71.

[36] Lu J, Rohani S. Preparation and characterization of theophylline– nicotinamide cocrystal. Organic Process Research & Development. 2009;13 (6) 1269-1275.

[37] Aakeröy CB, Fasulo ME, Desper J. Cocrystal or salt: does it really matter? Molecular Pharmaceutics. 2007;4 (3):317-22.

[38] Basavoju S, Boström D, Velaga SP. Indomethacin–saccharin cocrystal: design, synthesis and preliminary pharmaceutical characterization. Pharmaceutical Research. 2008;25 (3) 530-541.

[39] Pinto Vitorino G, Sperandeo NR, Caira MR, Mazzieri MR. A supramolecular assembly formed by heteroassociation of ciprofloxacin and norfloxacin in the solid state: co-crystal synthesis and characterization. Crystal Growth & Design. 2013;13 (3) 1050-1058.

[40] Velaga SP, Basavoju S, Boström D. Norfloxacin saccharinate–saccharin dihydrate cocrystal–A new pharmaceutical cocrystal with an organic counter ion. Journal of Molecular Structure. 2008;889 (1) 150-153.

[41] Johnson S, Rumon K. Infrared spectra of solid 1: 1 pyridine-benzoic acid complexes; the nature of the hydrogen bond as a function of the acid-base levels in the complex1. The Journal of Physical Chemistry. 1965;69 (1) 74-86.

[42] Jung MS, Kim JS, Kim MS, Alhalaweh A, Cho W, Hwang SJ, et al. Bioavailability of indomethacin-saccharin cocrystals. Journal of Pharmacy and Pharmacology. 2010;62 (11) 1560-1568.

[43] Padrela L, Rodrigues MA, Velaga SP, Matos HA, de Azevedo EG. Formation of indomethacin–saccharin cocrystals using supercritical fluid technology. European Journal of Pharmaceutical Sciences. 2009;38 (1) 9-17.

[44] Vogt FG, Clawson JS, Strohmeier M, Edwards AJ, Pham TN, Watson SA. Solid-state NMR analysis of organic cocrystals and complexes. Crystal Growth and Design. 2008;9 (2) 921-937

[45] Chadha, R.; Saini, A.; Arora, P.; Bhandari, S. Pharmaceutical cocrystals: a novel approach for oral bioavailability enhancement of drugs. Critical Reviews™ *in Therapeutic* Drug Carrier Systems.2012, 29, 183–218.

[46] Smith AJ, Kavuru P, Wojtas L, Zaworotko MJ, Shytle RD. Cocrystals of quercetin with improved solubility and oral bioavailability. Molecular Pharmaceutics. 2011;8 (5):1867–1876.

[47] Smith AJ, Kavuru P, Arora KK, Kesani S, Tan J, Zaworotko MJ, et al. Crystal engineering of green tea epigallocatechin-3-gallate (EGCg) cocrystals and pharmacokinetic modulation in rats. Molecular Pharmaceutics. 2013;10 (8):2948-2961.

[48] Timmons DJ, Bernas AA, Pacheco MR, Fricke KA. Crystal engineering with flavonoids; The 235th ACS National Meeting, New Orleans, LA, April 6–10, 2008; http://oasys2.confex.com/acs/235nm/techprogram/P1163554.HTM

[49] Pujari TA. Cocrystals of nutraceuticals: Protocatechuic acid and quercetin, 2009. Theses and Dissertations. Paper 2156; http://scholarcommons.usf.edu/etd/2156

[50] Kavuru P. Crystal engineering of flavonoids, Master of Science, Department of Chemistry, College of Arts and Sciences, University of South Florida; http://purl.fcla.edu/usf/dc/et/SFE0002463 http://scholarcommons.usf.edu/etd/325/

[51] Smith AJ, Kavuru P, Wojtas L, Zaworotko MJ, Shytle RD. Cocrystals of quercetin with improved solubility and oral bioavailability. Molecular Pharmaceutics. 2011;8 (5):1867–1876.

[52] Jin GZ, Yamagata Y, Tomita K. Structure of naringin hexahemihydrate. Acta Crystallographica. 1990;C46:310–313

[53] Schultheiss NC, Bethune SJ. Pterostilbene cocrystals. United States Patent Application 20110189277, 08/04/2011; http://www.freepatentsonline.com/y2011/0189277.html,Schultheiss N, Bethune S, Henck JO. Nutraceutical cocrystals: utilizing pterostilbene as a cocrystal former. Crystal Enginnering Communication 2010;12:2436–2442.

[54] Schultheiss N, Bethune S, Henck JO. Nutraceutical cocrystals: utilizing pterostilbene as a cocrystal former. Crystal Engineering Communication. 2010;12 2436–2442.

[55] Bethune SJ, Schultheiss N, Henck JO. Improving the poor aqueous solubility of nutraceutical compound pterostilbene through cocrystal formation. Crystal Growth & Design. 2011;11 (7) 2817–2823.

[56] Schultheisis N, Roe M, Boerrigter Stephan XM. Cocrystals of nutraceutical p-coumaric acid with caffeine and theophylline: polymorphism and solid-state stability explored in detail using their crystal graphs. Crystal Enginnering Communication 2011, 13 (2) 611-619.

[57] Melanie J. Bevill *, Petinka I. Vlahova, and Jared P. Smit. Polymorphic Cocrystals of Nutraceutical Compound *p*-Coumaric Acid with Nicotinamide: Characterization, Relative Solid-State Stability, and Conversion to Alternate Stoichiometries.

[58] Sanphui P, Goud NR, Khandavilli UBR, Nangia A. Fast dissolving curcumin cocrystals. Crystal Growth & Design. 2011;11 (9) 4135–4145.

[59] Elbagerma MA, Edwards HGM, Munshi T, Scowen IJ. Identification of a new cocrystal of citric acid and paracetamol of pharmaceutical relevance. Crystal Engineering Communication. 2011;13 1877–1884.

[60] Aboarayes DA. Crystal engineering of nutraceutical cocrystals: University of South Florida; 1820.

[61] Wang S, Chen J. Gossypol co-crystals and the use there of. 03/26/2009, Patent application number:20090082445.

[62] Sowa M, Ślepokura K, Matczak-Jon E. Cocrystals of fisetin, luteolin and genistein with pyridinecarboxamide coformers: crystal structures, analysis of intermolecular interactions, spectral and thermal characterization. Crystal Engineering Communication. 2013;15 (38):7696-708.

[63] Sowa M, Ślepokura K, Matczak-Jon E. Solid-state characterization and solubility of a genistein-caffeine cocrystal. Journal of Molecular Structure. 2014; 1076: 80–88

Chapter 6

Recrystallization of Drugs — Effect on Dissolution Rate

Yousef Javadzadeh, Solmaz Maleki Dizaj,
Zhila Vazifehasl and Masumeh Mokhtarpour

Additional information is available at the end of the chapter

http://dx.doi.org/10.5772/60006

1. Introduction

Undoubtedly, a broad range of beneficial pharmaceutical agents available today is one of the greatest scientific achievements. In recent years, numerous chemical and pharmaceutical researches have been led to the production of a vast number of new potential drug candidates. However, many of these new compounds are in a biopharmaceutical classification of low solubility keeping drug dissolution rate as the limiting step for absorption of these drugs [1]. Many efforts are being performed to reformulation of present drugs to solve problems such as low solubility, dissolution, bioavailability, flowability, compressibility, dosing problem, stability and toxicity. Appropriate drug dissolution rate is necessary for attaining proper oral drug therapy. The bioavailability of orally administrated drugs depends on their absorption from the gastrointestinal (GI) tract [2]. In this rout of administration, poorly water soluble drugs show low bioavailability because of their low solubility in GI media. The rate limiting step in the absorption of these drugs is the dissolution rate of them in the GI fluids rather than their diffusion through the GI membrane [3]. There are several techniques have been used for enhancing dissolution rate of these drugs such as: physical and chemical modification, preparing solid dispersion, complexation, solubilization, liquisolid technique, changing in crystal habit and so on [4].

As we know, certain kinds of drugs might be existed in a variety of solid-state forms, including polymorphs, solvates, hydrates, salts, cocrystals and amorphous solids. Then, selection of the proper form of the drug would be severely important to a drug development processes. Crystallization of a drug is an important thermophysical process in pharmaceutics that a drug molecule precipitates in the medium due to supersaturation [5]. In general, crystallization may be defined as the phenomena in which a solid compound precipitates from a saturated solution (through cooling or evaporation) in the form of crystals [6]. Enhancement of dissolution rates of water insoluble drugs (Sulfathiazole, Prednisone and Chloramphenicol) by crystallization has been investigated by Chiou et al in 1976 [7].

According to numerous papers, crystal habit can be modified by recrystallizing the drug. This process can affect the physical and physicochemical properties. Variation in crystalline habit is one of the proceeding trends in order to increasing the solubility, dissolution rate and bioavailability of the poorly soluble drugs [8]. This process could potentially be utilized to a wide range of drugs with different crystalline forms. Different crystals show different dissolution rates and then different biological responses. Therefore the modification of crystal properties should affect the bulk dissolution rate. Burt and Nickel (1979) by using the single crystal dissolution method demonstrated that the different forms of the same crystals prove different dissolve rates. Habit modification normally occurs when the environments of growing crystals have effects on its external shape without changing its internal structure [9]. Kobayashi et al (2000) reported different dissolution rates for carbamazepine, where the dihydrate form of the drug in simulated fluids (pH 1.2) had notably slower dissolution rates than the anhydrous forms (forms I and III). Due to the rapid conversion of metastable polymorph (form III) to the dehydrate form, this polymorph showed greatest rates of dissolution at the initial stages but reductions at later time [10].

2. Dissolution rate of drug

As previously mentioned, the dissolution and release rate of a drug molecule are essential to identification of drug characterization. The drug dissolution rate is an amount of a solid dosage form that goes into solution in a certain amount of time. In the dissolution test, the drug solubility is the main parameter that affects the rate of dissolution [11]. The poor dissolution rates of poorly water soluble drugs are still a challenge in the pharmaceutical industry [12, 13]. The absorption rate of a poorly water soluble drug in an orally administered solid dosage form is controlled by its dissolution rate in the fluid present at the absorption site and the dissolution rate is often the rate-determining step in its absorption [14].

The size and shape have very important role in the separation process which will in turn affect the yield and quality of the resulting fractions during the crystallization process [15]. Crystal habit influences particle orientation, therefore can modify the flowability, packing, compatibility, syringability, physical stability and dissolution profile of a drug molecule. For example, it has been showed that symmetrically shaped crystals of Ibuprofen have better compaction and flow properties than needle shaped crystals [16]. Therefore optimization of crystal properties is an alternative method for modifying the dissolution properties of drugs and therefore bioavailability of them [17].

There are many various factors that affect the dissolution rate of a drug substance. Some of these factors include: concentration of the drug solution, size and surface area of the drug substance, any movement or stirring, temperature, nature and type of drug matter and the mass or volume of the drug [8]. By breaking a drug into smaller pieces, its surface area is increased. When particle becomes smaller, because of greater interaction with the solvent, the solubility will be increased [18]. According to Noyes-Whitney equation, when particle size is reduced, the total effective surface area is increased and thereby the dissolution rate is enhanced. Indeed, when the total surface area of the drug particles is increased, the drug dissolves more rapidly because the action takes place only at the surface of each particle and then increases its dissolution rate [19-21]. Rising of temperature for liquids and solid solutes, not only increases the amount of the dissolving solute but also can enhanced the dissolving rate [22]. When the amount of solute in the solution is low, dissolution occurs relatively rapidly. In the point that no solute can be dissolved, dissolution takes place more slowly. In the case of liquid and solid solutes, stirring leads to fresh portions of the solvent in contact with the solute, then the increased dissolution rate was obtained [23].

So far many techniques have been used to increase the solubility and therefore the dissolution rate of poorly water soluble drugs. Methods such as cosolvency and microemulsion techniques [24-27], crystal habit modification, complexation [28-31], formulation in proper salt form [32, 33], preparation of solid dispersion [8, 34-39], liquisolid technique [19, 40-44] and particle size reduction [45-48] are some of these techniques. There are several disadvantages in the use of each technique. For example, the salt formation process is complex procedure and the dissolution enhancement is not always predictable in this process. Furthermore, this method is not possible for neutral compounds [49]. As another example, solubilization techniques normally show poor stability and insufficient acceptability of patients [50]. For choosing a suitable technique for a certain drug, some certain aspects such as the properties of drug, nature of excipients and nature of dosage form should be considered [51]. Crystal habit of a solid stated form can be modified by recrystallizing process. Then this process can affect the physical and physicochemical properties such as the melting point, solubility, true density, dissolution profile, flowability and tabletability [52].

3. Recrystallization of drugs

Most active pharmaceutical ingredients are manufactured in the crystalline form for their chemical stability during transportation, packaging and storage. Recrystallization is the simple and inexpensive method for purifying solid organic materials whereby a crystalline form of a compound may be obtained from other solid-state forms of the same substance [53]. In the pharmaceutics, recrystallization is the process most often applied for the purification of solid active pharmaceutical ingredients (APIs) [54]. Indeed, production of another crystalline form as well as purification are the two main aspects in the recrystallization process of drugs. Generally, a hot saturated solution of the drug is prepared using a solvent with appropriate boiling point and then the solution is cooled and a new crystalline form of the drug is separated [53]. This process occurs due to the lower solubility of new crystalline form of the drug in the relevant solvent at lower temperatures. The recrystallization process of a drug has been schematically illustrated in figure 1.

Recrystallization of the drug in a suitable solvent leading to altering physical and chemical properties of obtained particles. These properties are consist of the wettability, melting point,

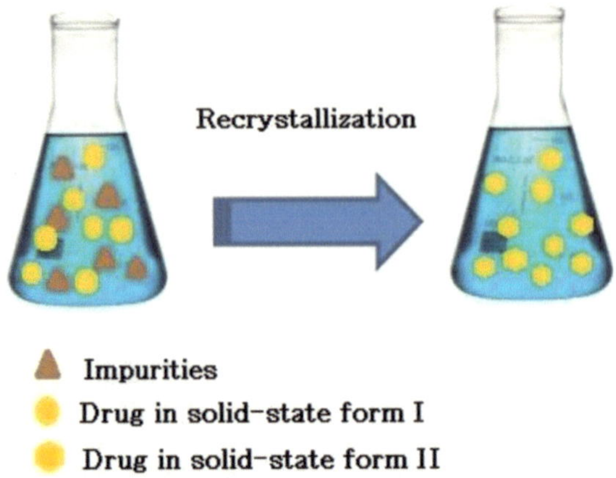

Figure 1. Schematic illustration of the drug recrystallization process.

solubility, dissolution rate, flowability and compactibility [55]. There are some reports about the recrystallization process and its effects on different properties of drugs [56-58]. This technique has been used for huge or small quantities of materials and very effective and efficient results were obtained.

Various factors including thermodynamic (*e.g.*, solubility, solid-liquid interfacial tension, solvent activity, temperature, *etc.*), kinetic (*e.g.*, supersaturation, molecular mobility, metastable zone width), and molecular recognition (hydrogen bonds, non-covalent bonds, molecular networks) may influence on the rate and mechanisms by which crystals are formed from liquid solutions [59, 60]. The driving force for crystallization is supersaturation. Supersaturation can be created by increasing the solute concentration (solvent evaporation) or decreasing the solute solubility (*e.g.*, temperature change, antisolvent addition, pH change, salting out) [61]. The next step is the formation of macroscopic crystals from stable nuclei, known as crystal growth. This step is controlled by internal (crystal structure) and external factors (temperature, impurities, supersaturation, solvent type) and determines the particle morphology [62].

It was shown that changing of crystal habit by recrystallization has significant effect on dissolution rate. Nogami and Kato found that the recrystallized thin crystal of the aspirin shows better dissolution and excretion behavior compare to the commercial aspirin [63].

Valsartan (VAL) is a potent, highly selective and orally active antihypertensive drug with sparingly solubility in aqueous media (especially in gastric fluids). Therefore its absorption is limited by its dissolution rate. Nalluri et al. prepared and investigated the recrystallized form of this drug. According to their results (evidenced by the DSC and powder X-RD studies), by applying of the recrystallization technique, VAL can be transposed into either amorphous phases or amorphous dispersions. The recrystallized VAL exhibited improved aqueous solubility and dissolution properties and potentially therapeutic efficacy compared to an untreated drug [52].

Nokhodchi et al. recrystallized phenytoin in ethanol and acetone. According to their results, needle-like and rhombic crystal habits exhibit different dissolution rates. They suggested that the differences in dissolution rates for phenytoin were related with the surface area of various crystals with different shapes [64].

Carbamazepine is a poorly water soluble drug and its absorption may be unpredictable. This drug is applied as first line monotherapy for seizures in elderly patients. Proven efficacy and less cost are two major advantages of carbamazepine. Mahalaxmi et al. recrystallized carbamazepine by using two solvents and at different cooling conditions. Their outcomes showed improved dissolution behavior of the treated drug [65].

Adhiyaman et al. reported different properties of dipyridamole crystals that prepared by the recrystallization process. Dipyridamole is a critical antiplatelet and peripheral vasodilator drug that has poor water solubility and then bioavailability, which limits its effectiveness in clinical tries. Their results showed that the prepared crystals of dipyridamole under optimized conditions, significantly improved the dissolution rate of the drug compared to untreated one [66].

Applying of solvents or high temperatures did not modify the crystalline form of ketoconazole. It was reported that adjusting of growth of crystal faces can affect on the physicochemical properties of the ketoconazole crystals. It was demonstrated that habit modification and different degrees of crystallinity are significant parameters in the enhancement of the dissolution rate [67].

3.1. Recrystallization steps

Choosing a proper solvent is the first step of the recrystallization process. After selection of a proper solvent, an impure compound is dissolved in the minimum amount of a hot solvent. Then solution is decolorized with activated charcoal (if needed) and any insoluble material is filtered off. If the crystal formation process occurs too rapidly, impurities may become trapped in the crystals. Slowly cooling of the resulting solution to induce crystallization is the next step. Finally the crystals are collected, washed and dried. Figure 2 shows steps of the recrystallization process [9, 66, 68].

3.1.1. Choosing a proper solvent

The most important feature of the recrystallization technique is the selection of the solvent. There are two main facts in purification of solid organic materials by the recrystallization method: different solids have different solubilities in a given solvent and most solids are more soluble in hot than in cold solvents [69, 70].

Numerous solvents are employed for recrystallization of drug compounds. Selection of the recrystallization solvent should be performed on a case-by-case basis [9, 68]. This is due to the variant solubility of different organic compounds in these solvents. It is clear that, at the first, the material should be introduced into solution and then can able to come back out of solution. Choosing the appropriate solvent is very important in recrystallization, because it leads to

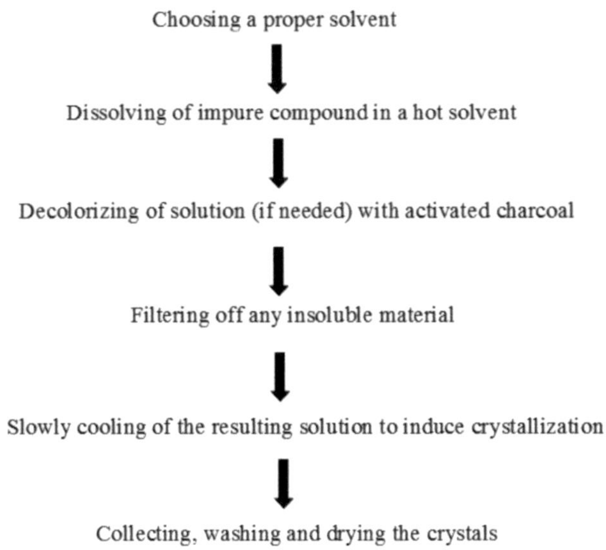

Figure 2. Steps of the recrystallization process.

increasing our yield of product [9, 25, 68]. This selected solvent should have a lower boiling point than the melting point of the solid that being recrystallized. Furthermore, the drug should be soluble in the boiling solvent but have poor solubility in the same solvent after cooling [25, 64].

3.1.2. Melting point of drug and solvent

A melting point is employed to identify a substance and to get an indication of its purity. The melting point of a solid is the temperature at which the vapor pressure of the solid phase becomes equal to the vapor pressure of the liquid phase. Purity of a recrystallized drug is evaluated by observing its color and by measuring its melting point range. According to literature, the boiling point of the recrystallization solvent should be lower than the melting point of the compound to be recrystallized. If it is higher than the melting point of drug, the drug will "oil out", the process in which a drug is insoluble in a solution at a temperature above its melting point [9, 66, 68].

3.2. Common procedures for recrystallization

Different polymorphs of active pharmaceutical ingredients can be obtained though recrystallization processes based on solvents, solid-state transitions and vapor deposition [54].

In the controlled kinetic or thermodynamic conditions, crystallization of drug polymorphs can be performed from a solvent system. Supercritical fluid techniques are used for micronization and recrystallization of different drugs. According to the reports, modification of solid state characteristics, such as crystal habit, crystallinity and polymorphism, has been successfully achieved through recrystallization of drug particles using various supercritical antisolvent (SAS) processes [70, 71]. This technology by using carbon dioxide gas is able to modify the solid state properties of drugs, such as characteristics of particles (size, shape, surface, crystal structure and morphology), crystallinity and polymorphism leading to affecting their dissolution rate and bioavailability [72]. The supercritical anti solvent technique is based on injection of an organic solution (containing of drug crystals) into supercritical carbon dioxide. During of the mixing, supercritical CO_2 is quickly dissolved in the organic solution and leads to the precipitation of solutes (drug) by an antisolvent effect. Then supercritical CO_2 efficiently extracts the organic solvent, leads to achieving completely solvent-free products [71]. Figure 3 shows recrystallization of a drug by using the supercritical antisolvent (SAS) process schematically.

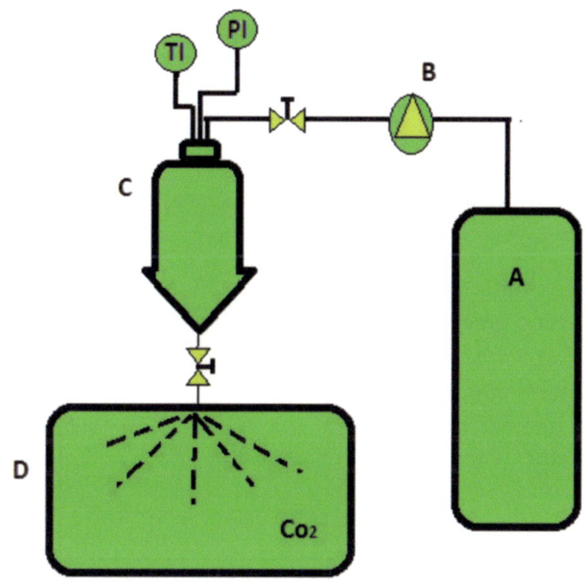

Figure 3. Recrystallization by using the supercritical antisolvent (SAS) process. A) Drug solution in good solvent; B) Peristaltic pump; C) Spraying gun for drug solution; D) Chamber with saturated CO_2 gas

Some drugs have also been recrystallized by the spherical agglomeration technique in order to modify their dissolution properties. For poorly water soluble drugs, a water immiscible organic solvent can act as an external phase and a 20 % calcium chloride as a bridging liquid [73]. The spherical crystallization is a particle size enlargement method that includes crystallization and agglomeration using bridging liquid [74]. Indeed, this method is a particle design technique that crystallization and agglomeration can be performed simultaneously in one step to transform crystals directly into the compacted spherical form [75, 76]. By using of this method, direct tabletting of drug instead of further processing like mixing, granulation, sieving and drying is possible [73, 76, 77]. In 1986, Kawashima applied the spherical crystallization method for size enlargement of drugs in the pharmaceutical field. Martino et al prepared spherical propyphenazone crystals by the agglomeration recrystallization technique using a three solvent system [78]. Figure 4 shows the mechanism of the recrystallized agglomerates formation by the spherical agglomeration method.

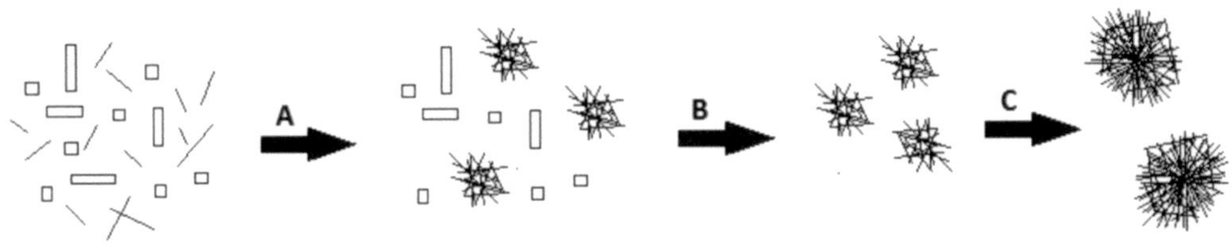

Figure 4. Mechanism of the recrystallized agglomerates formation by the spherical agglomeration technique; A) Recrystallization by addition of bridging liquid, B) Random coalescence, C) Enlarged spherical agglomerates.

The most commonly used spherical crystallization techniques are spherical agglomeration and quasi-emulsion solvent diffusion. In both processes, a good solvent that dissolves the compound to be crystallized is used. A poor solvent (as an antisolvent) is used to generate the required supersaturation [74, 77].

Polymorphs may also be generated through solid-state transition by recrystallization of metastable polymorphs and amorphous forms, or through desolvation of solvates (including hydrates) [60]. Physical vapor deposition (PVD) method is an atomistic deposition procedure in which material is vaporized from a solid or liquid source in the form of atoms or molecules and transported in the form of vapor through a low pressure environment to the substrate where it condenses [60, 79].

4. Mechanism of recrystallization that effects on dissolution rate

In the nature, crystals can be categorized into seven crystal systems with various properties. Each of these crystal systems has differences in the relative sizes of their faces as well. This variation is called crystal habit modifications. In fact the morphology of a crystal depends on the growth rates of the different faces of it. Fast growing faces have little or no effect on the growth form while slow growing faces have more influence. Environmental conditions and structure and defects of crystal governed the growth of a given face [80]. The interaction of the solvent at the different crystal solution interfaces may cause to modified roundness of growing crystal faces (or edges), changes in crystal growth kinetics and enhancement or inhibition of crystal growth at certain faces [54]. Figure 5 shows the crystal habits arising from growth inhibition at crystal faces.

http://dx.doi.org/10.5772/60006

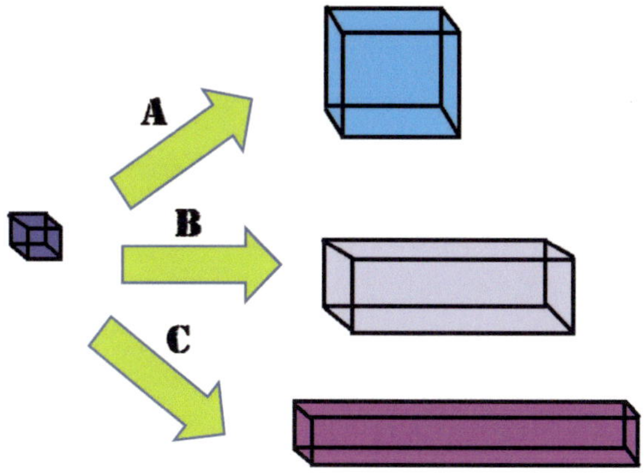

Figure 5. Crystal habits arising from growth inhibition at crystal faces; A) Growth equal in all dimensions, B) Growth inhibited in one dimension, C) Growth inhibited in two dimensions.

In pharmaceutical manufacturing, a crystal habit of a drug molecule is an important characteristic. Different crystals have different planes, specific surface and free surface energy. Therefore they show different physicochemical properties such as dissolution rate, flowability and compressibility [68]. The existence of the drug structure in various crystalline shape influence on the bioavailability and biological efficacy of a drug [66]. The crystalline form that has the improved dissolution rate and solubility is useful for developing the efficient pharmaceutical product of drugs [68, 81].

The primary mechanism of recrystallization is the long-range motion of grain boundaries that removes dislocations from the material [82]. According to numerous reports wettability and consequent dissolution rate of different crystalline forms of drugs could be determined by exposure of diverse crystal faces [83]. The constituent molecules of a crystalline material are part of an orderly three-dimensional structure [84]. Depending on the crystallization conditions, these molecules arrange in different forms. This crystalline form of material is known as polymorphic and each of its crystal ordering is called a polymorph. Indeed, the polymorphs are substances with the same chemical composition but presenting different crystalline forms [85]. Different crystalline forms include various crystallographic systems, lattice parameters and position of the atoms in the elementary cell [86]. Under different conditions, a crystalline drug is crystallized into different polymorphic forms. Polymorphs of a substance show different physical properties, such as solubility and dissolution properties. In fact, crystals with higher free energy show higher solubility. Since the molecules in crystals with high energy have weaker solid-solid interactions compared to crystals with lower energy, they exhibit higher affinity for the solvent in the surroundings [87]. Figure 6 shows the probable consequences of a different crystal form in solubility.

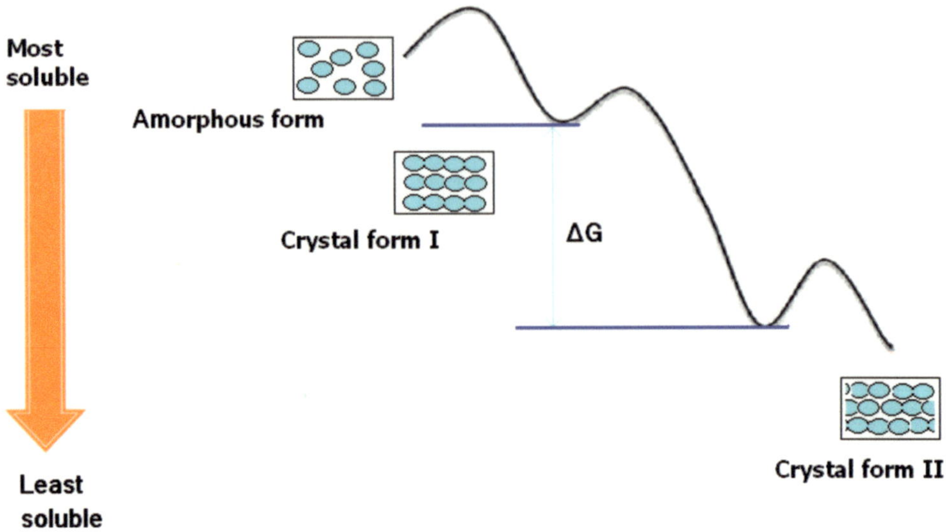

Figure 6. Probable consequences of a different crystal form in solubility

Figure 7 show formation of different crystals following by recrystallization of the initial crystalline form schematically. First, the reaction between solid and solvent breaks the solid crystalline substances (opens the amorphous lattice for amorphous substances). This creates cavities in the solvent (known as phase change) and, hence, molecules of solid become molecules of solute. Then solute molecules recrystallized into another crystalline form(s) [88].

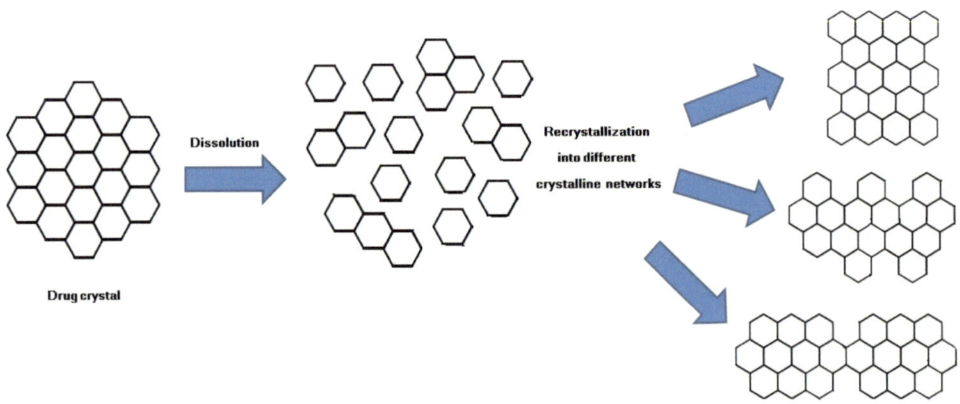

Figure 7. First the reaction between solid and solvent breaks the solid crystal for crystalline substances (opens the amorphous lattice for amorphous substances). This creates cavities in the solvent (known as phase change) and, hence, molecules of solid become molecules of solute. Then solute molecules recrystallized into another crystalline form(s).

Additional information is necessary for control of the nucleation and crystal growth. Dunitz and Bernstein have evaluated the disappearing polymorphs by considering various examples. They state that "once a particular polymorph has been obtained, it is always possible to obtain it again; it is only a matter of finding the right experimental conditions" [89].

5. Effect of excipients and additives on the recrystallization process

Crystal morphology engineering is a useful tool for improving tabletting efficiency of pharmaceutical solids. Pharmaceutically accepted excipients can be applied as crystal habit modifiers to overcome the toxicity and/or environmental concerns [90]. Excipients and polymers have a major role to prepare the recrystallized form of a drug. The difference in the physicochemical properties of the drug molecules and the excipient is the main factor in selection of a solvent system for the recrystallization process. The presence of additives such as the polymeric material and surface active agents can influence molecular aggregation during crystallization. Viscosity of the medium and surface tension are reduced by the surfactants that affect the nucleation process. The presence of these additives in the spherical agglomeration process may also reduce the processing time and improves the bioavailability and the micrometric properties of drug [91].

Various polymers like hydroxy propyl methylcellulose (HPMC), poly ethylene glycol (PEG), ethyl cellulose (EC) and poly vinyl pyruvate (PVP) can be used to improve poor compressibility and handling qualities of pure drugs by crystallization techniques [92]. This improves the micromeritic and drug release properties of the crystals as well.

Crystallization is inhibited by some of the polymers such as methylcellulose, Hyroxypropyl methyl celluose (HPMC), Polyvinylpyrrolidone (PVP). Among these, PVP has been found to be the most effective in inhibiting crystallization of drugs. These antinucleant polymers are incompatible in both size and size of the host molecules of the growing crystals surface. Then, their incorporation into the lattice alters growth characteristics of the host molecules [93]. Recrystallization of the amorphous drug may be avoided by entry of stabilizing hydrophilic excipients in the composition. Indeed, stabilizing excipients are prevented recrystallization of amorphous drug particles by steric hindrance and/or the formation of hydrogen bond with drug molecules [94]. As a noticeable point, the stabilizing excipients should be hydrophilic for improving the wetting properties of drug particles and these hydrophilic excipients should not be hygroscopic as well.

6. Methods for characterization of crystals

There are different analytical methods and techniques for characterization of various properties of crystals and therefore efficiency of the recrystallization procedure. Measuring of size, crystal size distribution and surface microtopographic observations can help to prediction of growth mechanisms of crystal faces under different conditions. Monitoring the surface topography during crystal growth can display events that are not evident from morphology studies. Some characterization techniques of the crystals are collected in table 1.

Technique(s)	Characteristic(s)
Sieve analysis, laser diffraction, and computerized image analysis.	Size and size distribution of crystals [95].
Optical microscope	Size, morphology and shape of crystals [96].
Scanning electron microscopy	Size, surface topography, polymorphism and crystal habit [95].
X-ray powder diffraction	It is important for establishing batch-to-batch reproducibility of a crystalline form. The type of crystal can be determined by using this technique. An amorphous form does not form a special pattern [97].
Fourier Transform Infrared spectrometer (FTIR)	It is much more beneficial for identifying between solvates and anhydrous form [95].
Differential scanning calorimeter (DSC)	It can measure the heat loss or gain resulting from physical or chemical changes within a sample. If a mixture of drugs and polymer is crystallized together then variation in properties of crystals can be evaluated with DSC [75].
Angle of repose, angle of spatula, angle of fall and Carr's index.	Flowability [98].
Contact angle of water to the compressed crystals	Wettability (As the contact angle is decreased the wettability is enhanced) [99].
By measuring the interfacial tension between the bridging liquid and the continuous phase, contact angle and the ratio of the volumes of the bridging liquid and solid particles.	The strength of the agglomerated crystals [78].
From the comparison of stress relaxations and elastic recovery the compressibility of the spherical and single crystals can be evaluated.	Compressibility [100].

Table 1. Characterization techniques of the crystals

Understanding the thermodynamic and kinetic behavior of the system is necessary for developing of reliable processes. There are various techniques that can be informed us about crystallization kinetics and growing mechanisms.

Various information can be obtained by using of an optical microscopy (inverted microscope); studying of crystallization processes *in situ*, monitoring of the transformations in suspensions, determination of transformation times, screening and characterization of additive/solvent interactions with specific crystal faces, identification of nucleation mechanisms and measuring the crystal growth rate [96]. Three latter mentioned points can be achieved by using an electron microscopy and atomic force microscopy techniques as well. Monitoring of molecular association processes that direct nucleation and crystal growth can be performed by Raman spectroscopy [95]. Spectrophotometry and chromatography techniques are applied for measuring of the concentration of solute in solution and supersaturation process, and a diffraction calorimetry spectroscopy for monitoring the solid phase composition of crystals [101, 102].

7. Conclusion

The solubility, dissolution rate and crystal properties of a drug substance play a key role in the pharmaceutical development, manufacturing and formulation. Low solubility of a certain crystalline form of a drug leads to the low dissolution rate and bioavailability and therefore reduced biological responses. Low solubility reduces also rates of drug clearances and then increases the toxicity of the drug product. Drug crystals can be modified by the recrystallization technique and this process can affect the physical and physicochemical properties such as melting point, solubility, true density, dissolution profile, flowability and tabletability. The recrystallization method is simple and inexpensive enough for scaling up to a commercial level. This method has showed an efficient role in the increasing of the bioavailability of poorly soluble drugs. Understanding the thermodynamic and kinetic behavior of the system is necessary for the development of suitable recrystallization processes. Under controlled conditions, recrystallization may lead to enhanced surface area and revelation of more polar surface moieties, therefore cause to enhanced dissolution rate. Conversion to a more stable polymorph, recrystallization of the amorphous material or reduced chemical stability are some of the recrystallization problems in regard to solubility and dissolution rate of drugs.

Author details

Yousef Javadzadeh*, Solmaz Maleki Dizaj, Zhila Vazifehasl and Masumeh Mokhtarpour

*Address all correspondence to: javadzadehy@yahoo.com

Drug Applied Research Center and Faculty of Pharmacy, Tabriz University of Medical Sciences, Iran

References

[1] Yasir, M., M. Asif, A. Kumar, and A. Aggarval, *Biopharmaceutical classification system: an account.* International Journal of PharmTech Research, 2010. 2(3): p. 1681-1690.

[2] Gavhane, Y.N. and A.V. Yadav, *Loss of orally administered drugs in GI tract.*Saudi Pharmaceutical Journal, 2012. 20(4): p. 331-344.

[3] Mudie, D.M., G.L. Amidon, and G.E. Amidon, *Physiological parameters for oral delivery and in vitro testing.* Mol Pharm, 2010. 7(5): p. 1388-405.

[4] Sharma, D., M. Soni, S. Kumar, and G. Gupta, *Solubility enhancement—eminent role in poorly soluble drugs.* Research Journal of Pharmacy and Technology, 2009. 2(2): p. 220-224.

[5] Kiani, H. and D.-W. Sun, *Water crystallization and its importance to freezing of foods: A review.* Trends in Food Science&Technology, 2011. 22(8): p. 407-426.

[6] Mandelkern, L., *Crystallization of polymers.* Vol. 38. 1964: McGraw-Hill New York.

[7] Chiou, W.L., S.J. Chen, and N. Athanikar, *Enhancement of dissolution rates of poorly water-soluble drugs by crystallization in aqueous surfactant solutions I: Sulfathiazole, prednisone, and chloramphenicol.* Journal of pharmaceutical sciences, 1976. 65(11): p. 1702-1704.

[8] Joshi, H.N., R.W. Tejwani, M. Davidovich, et al., *Bioavailability enhancement of a poorly water-soluble drugby solid dispersion in polyethylene glycol–polysorbate 80 mixture.* International Journal of Pharmaceutics, 2004. 269(1): p. 251-258.

[9] Keraliya, R.A., T.G. Soni, V.T. Thakkar, and T.R. Gandhi, *Effect of solvent on crystal habit and dissolution behaviorof tolbutamide by initial solvent screening.* Dissolution Technol, 2010. 1: p. 16-21.

[10] Kobayashi, Y., S. Ito, S. Itai, and K. Yamamoto, *Physicochemical properties and bioavailability of carbamazepine polymorphs and dihydrate.* International journal ofpharmaceutics, 2000. 193(2): p. 137-146.

[11] Banakar, U.V., *Pharmaceutical dissolution testing.* Drugs and the pharmaceutical sciences, 1991. 49: p. 1-426.

[12] Kottke, M.K. and E.M. Rudnic, *Tablet dosage forms*2002: Dekker, New York, NY.

[13] Alderborn, G., *Tablets and compaction.* 2001.

[14] Singh, B.N. and K.H. Kim, *Drug delivery-oral route.* Encyclopedia of pharmaceutical technology, 2002: p. 886-909.

[15] Normah, I., C. Cheow, and C. Chong, *Crystal habit during crystallization of palm oil: effect of time and temperature.* International Food Research Journal, 2013. 20(1): p. 417-422.

[16] Prathipati, S. and V. Ganesan, *Spherical crystallization: a method to improve physicochemical properties.* Int. J Pharm Sci Review and Res, 2011. 6(1): p. 60-63.

[17] Javadzadeh, Y., A. Mohammadi, N.S. Khoei, and A. Nokhodchi, *Improvement of physicomechanical properties of carbamazepine by recrystallization at different pH values.* Acta pharmaceutica, 2009. 59(2): p. 187-197.

[18] Larsson, J., A. Axelsson, and M. Marucci, *Methods for measurement of solubility and dissolution rate of sparingly soluble drugs*, in *Department of Chemical Engineering, Faculty of Engineering*2009 Lund University.

[19] Javadzadeh, Y., L. Musaalrezaei, and A. Nokhodchi, *Liquisolid technique as a newapproach to sustain propranolol hydrochloride release from tablet matrices.* International journal of pharmaceutics, 2008. 362(1): p. 102-108.

[20] Atashafrooz, Z., S. Maleki Dizaj, and A. Salehi Sadaghiani, *Cucurbita pepo oil as a drug microemulsion formulation: study of phase diagram.* Nanomedicine Journal, 2014. 1(5): p. 298-301.

[21] Dokoumetzidis, A. and P. Macheras, *A century of dissolution research: from Noyes and Whitney to the biopharmaceutics classification system.* International journal of pharmaceutics, 2006. 321(1): p. 1-11.

[22] Sarkari, M., J. Brown, X. Chen, et al., *Enhanced drug dissolution using evaporative precipitation into aqueous solution.* International journal of pharmaceutics, 2002. 243(1): p. 17-31.

[23] Aulton, M.E. and T. Wells, *Pharmaceutics: The science of dosage form design.* 2002.

[24] Raval, A.J. and M.M. Patel, *Preparation and characterization of nanoparticles for solubility and dissolution rate enhancement of meloxicam.* Int Res J Pharm, 2011. 1(2): p. 83-89.

[25] Pandya, P., S. Gattani, P. Jain, L. Khirwal, and S. Surana, *Co-solvent evaporation method for enhancement of solubility and dissolution rate of poorly aqueous soluble drug simvastatin: in vitro–in vivo evaluation.* AAPS PharmSciTech, 2008. 9(4): p. 1247-1252.

[26] Maleki Dizaj, S., *Preparation and study of vitamin A palmitate microemulsion drug delivery system and investigation of co-surfactant effect,.* Journal Of Nanostructure in Chemistry, 2013. 2013.

[27] Nokhodchi, A., R. Aliakbar, S. Desai, and Y. Javadzadeh, *Liquisolid compacts: the effect of cosolvent and HPMC on theophylline release.* Colloids Surf B Biointerfaces, 2010. 79(1): p. 262-9.

[28] Jagdale, S.C., V.N. Jadhav, A.R. Chabukswar, and B.S. Kuchekar, *Solubility enhancement, physicochemical characterization and formulation of fast-dissolving tablet of nifedipine-betacyclodextrin complexes.* Brazilian Journal of Pharmaceutical Sciences, 2012. 48(1): p. 131-145.

[29] Dhanaraju, M.D., K.S. Kumaran, T. Baskaran, and M.S. Moorthy, *Enhancement of bioavailabilityof griseofulvin by its complexation with beta-cyclodextrin.* Drug Dev Ind Pharm, 1998. 24(6): p. 583-7.

[30] Shulman, M., M. Cohen, A. Soto-Gutierrez, et al., *Enhancement of naringenin bioavailability by complexation with hydroxypropoyl-β-cyclodextrin.* PloS one, 2011. 6(4): p. e18033.

[31] Sabzevari, A., K. Adibkia, H. Hashemi, et al., *Improved anti-inflammatory effects in rabbit eye model using biodegradable poly beta-amino ester nanoparticles of triamcinolone acetonide.* Invest Ophthalmol Vis Sci, 2013. 54(8): p. 5520-6.

[32] Patel, A., S.A. Jones, A. Ferro, and N. Patel, *Pharmaceutical salts: a formulation trick or a clinical conundrum.* Br J Cardiol, 2009. 16: p. 281-6.

[33] Shukla, V., *Techniques for solubility enhancement of poorly soluble drugs: an overview.* Journal of Medical Pharmaceutical and Allied Sciences, 2012. 1: p. 18-38.

[34] Ajay, S., D. Harita, M. Tarique, and P. Amin, *Solubility and Dissolution Rate Enhancement of Curcumin Using Kollidon VA64 by Solid Dispersion Technique.* International Journal of PharmTech Research, 2012. 4(3): p. 1055-1064.

[35] Ismail, S., M. El-Mahdy, and S. Al-Kubati, *Enhancement of solubility and dissolution of nimesulide using solubilization, solid dispersion and complexation techniques.* Bull. Pharm. Sci., Assiut University, 2009. 32: p. 321-338.

[36] Kakran, M., N. Sahoo, and L. Li, *Dissolution enhancement of quercetin through nanofabrication, complexation, and solid dispersion.* Colloids and Surfaces B: Biointerfaces, 2011. 88(1): p. 121-130.

[37] Kawakami, K., *Modification of physicochemical characteristics of active pharmaceutical ingredients and application of supersaturatable dosage forms for improving bioavailability of poorly absorbed drugs.* Advanced drug delivery reviews, 2012. 64(6): p. 480-495.

[38] Ramakrishna, G.S., K. AnjiReddy, B.J.R.D. P, P.R. Rao, and P.R. Bhavanam, *Enhancement dissolution rate of iboprofen by solid dispersion method.* Journal of Innovative trends in Pharmaceutical Sciences, 2010. 1(4).

[39] Roul, L., N. Manna, R. Parhi, S. Sahoo, and P. Suresh, *Dissolution rate enhancement of Alprazolam by Solid Dispersion.* Indian Journal of Pharmaceutical Education and Research, 2012. 46(1): p. 41.

[40] Javadzadeh, Y., B. Jafari-Navimipour, and A. Nokhodchi, *Liquisolid technique for dissolution rate enhancement of a high dose water-insoluble drug (carbamazepine).* International Journal of Pharmaceutics, 2007. 341(1): p. 26-34.

[41] Rajesh, K., R. Rajalakshmi, J. Umamaheswari, and C.A. Kumar, *Liquisolid technique a novel approach to enhance solubility and bioavailability.* International Journal, 2011. 2229: p. 7499.

[42] Thadkala, K., S. Voruganti, and S.R. Badari, *Enhancement of Dissolution Rate of Nimesulide by Liquisolid Compaction Technique.* International Journal of PharmTech Research, 2012. 4(3): p. 1294-1302.

[43] Javadzadeha, Y., M. Siahi-Shadbada, M. Barzegar-Jalalia, and A. Nokhodchia, *Enhancement of dissolution rate of piroxicam using liquisolid compacts.* II FARMACO, 2005 60(4): p. 361–365.

[44] Nokhodchi, A., Y. Javadzadeh, M. Siahi-Shadbad, and M. Barzegar-Jalali, *The effect of type and concentration of vehicles on the dissolution rate of a poorly soluble drug (indomethacin) from liquisolid compacts.* J Pharm Pharm Sci, 2005. 12;8(1): p. 8-25.

[45] Liversidge, G.G. and K.C. Cundy, *Particlesize reduction for improvement of oral bioavailability of hydrophobic drugs: I. Absolute oral bioavailability of nanocrystalline danazol in beagle dogs.* International Journal of Pharmaceutics, 1995. 125(1): p. 91-97.

[46] Möschwitzer, J., *Particle Size Reduction Technologies in the Pharmaceutical Development Process.* American Pharmaceutical Review, 2010. 2010: p. 54-59.

[47] Mohammadi, G., A. Nokhodchi, M. Barzegar-Jalali, et al., *Physicochemical and anti-bacterial performance characterization of clarithromycin nanoparticles as colloidal drug delivery system.* Colloids and Surfaces B: Biointerfaces, 2011. 88(1): p. 39-44.

[48] Javadzadeh, Y., F. Ahadi, S. Davaran, et al., *Preparation and physicochemical characterization of naproxen-PLGA nanoparticles.* Colloids Surf B Biointerfaces, 2010. 1;81(2): p. 498-502.

[49] Sahu, B.P. and M.K. Das, *Nanosuspension for enhancement of oral bioavailability of felodipine.* Applied Nanoscience, 2013.

[50] Das, M.K., *Nanosuspension for enhancement of oral bioavailability offelodipine.* Appl Nanosci, 2014. 4: p. 189–197.

[51] Kesisoglou, F., S. Panmai, and Y. Wu, *Nanosizing — Oral formulation development and biopharmaceutical evaluation.* Advanced drug delivery reviews, 2007. 59(7): p. 631-644.

[52] Nalluri, B.N., M.R. Krishna, T. Rao, and P.A. Crooks, *Effect of Recrystallization on the Pharmaceutical Properties of Valsartan for Improved Therapeutic Efficacy.* Journal of Applied Pharmaceutical Science, 2012. 2(10): p. 126.

[53] Javadzadeh, Y., S. Hamedeyazdan, and S. Asnaashari, *Recrystallization of Drugs: Significance on Pharmaceutical Processing,* K. Sztwiertnia, Editor 2012.

[54] Stieger, N., W. Liebenberg, and M. Andreeta, *Recrystallization of Active Pharmaceutical Ingredients.* Crystallization: Science and Technology; Andreeta, MRB, Ed.; InTech: Rijeka, Croatia, 2012.

[55] Kovačič, B., F. Vrečer, and O. Planinšek, *Spherical crystallization of drugs.* Acta pharmaceutica, 2012. 62(1): p. 1-14.

[56] Shekunov, B.Y. and P. York, *Crystallization processes in pharmaceutical technology and drug delivery design.* Journal of crystal growth, 2000. 211(1): p. 122-136.

[57] Huang, L.-F. and W.-Q.T. Tong, *Impact of solid state properties on developability assessment of drug candidates.* Advanced Drug Delivery Reviews, 2004. 56(3): p. 321-334.

[58] Shibata, Y., M. Fujii, A. Suzuki, et al., *Effect of storage conditions on the recrystallization of drugs in solid dispersions with crospovidone.* Pharm Dev Technol, 2014. 19(4): p. 468-74.

[59] Mullin, J.W., *Crystallization*2001: Butterworth-Heinemann.

[60] Rodríguez-hornedo, N. and D. Murphy, *Significance of controlling crystallization mechanisms and kinetics in pharmaceutical systems.* Journal of pharmaceutical sciences, 1999. 88(7): p. 651-660.

[61] Chaudhary, A., U. Nagaich, N. Gulati, et al., *Enhancement of solubilization and bioavailability of poorly soluble drugs by physical and chemical modifications: A recent review.* Journal of Advanced Pharmacy Education & Research, 2012. 2(1): p. 32-67.

[62] Vedantam, S. and V.V. Ranade, *Crystallization: Key thermodynamic, kinetic and hydrodynamic aspects.* Sadhana, 2013. 38(6): p. 1287-1337.

[63] Yadhav, A., *Significance of recrystalization on pharmaceutical dosage form processing.* Latest Reviews, 2007. 3(3).

[64] Nokhodchi, A., N. Bolourtchian, and R. Dinarvand, *Crystal modification of phenytoin using different solvents and crystallization conditions.* International journal of pharmaceutics, 2003. 250(1): p. 85-97.

[65] Mahalaxmi, R., P.S. Ravikumar, A. Shirwaikar, and A. Shirwakikar, *Effect of recrystallization on size, shape, polymorph and dissolution of carbamazepine.* Int J Pharm Tech Res, 2009. 1(3): p. 725-732.

[66] Adhiyaman, R. and S.K. Basu, *Crystal modification of dipyridamole using different solvents and crystallization conditions.* International journal of pharmaceutics, 2006. 321(1): p. 27-34.

[67] Viseras, C., I.I. Salem, I. Galan, A.C. Galan, and A.L. Galindo, *The effect of recrystallization on the crystal growth, melting point and solubility of ketoconazole.* Thermochimica acta, 1995. 268: p. 143-151.

[68] Assaf, S.M., M.S. Khanfar, R. Obeidat, M. Sheikh-Salem, and A.I. Arida, *Effect of Different Organic Solvents on Crystal Habit of Mefenamic Acid.* Jordan Journal of Pharmaceutical Sciences, 2010. 2(2).

[69] Columbus, P.S., *Hot and cold water redispersible polyvinyl acetate adhesives*, 1981, Google Patents.

[70] Chrastil, J., *Solubility of solids and liquids in supercritical gases.* The Journal of Physical Chemistry, 1982. 86(15): p. 3016-3021.

[71] Kim, M.-S., S. Lee, J.-S. Park, J.-S. Woo, and S.-J. Hwang, *Micronization of cilostazol using supercritical antisolvent (SAS) process: Effect of process parameters.* Powder Technology, 2007. 177(2): p. 64-70.

[72] Van Ginneken, L. and H. Weyten, *Particle formation using supercritical carbon dioxide*, in*Carbon Dioxide Recovery and Utilization*2003, Springer. p. 123-136.

[73] Mahanty, S., J. Sruti, C.N. Patra, and M.B. Rao, *Particle design of drugs by spherical crystallization techniques.* Int. J. Pharm. Sci. Nanotech, 2010. 3: p. 912-918.

[74] Gupta, M., B. Srivastava, M. Sharma, and V. Arya, *Spherical crystallization: A tool of particle engineering for making drug powder suitable for direct compression.* Int J Pharm Res Devt, 2010. 1: p. 1-10.

[75] Parida, R., *Evaluation parameters for spherical agglomerates formed by spherical crystallization technique.* Int J Pharm Biol Sci, 2010. 1: p. 1-10.

[76] Paradkar, A., A. Pawar, J. Chordiya, V. Patil, and A. Ketkar, *Spherical crystallization of celecoxib.* Drug development and industrial pharmacy, 2002. 28(10): p. 1213-1220.

[77] Bharti, N., N. Bhandari, P. Sharma, K. Singh, and A. Kumar, *Spherical Crystallization: A Novel Drug Delivery Approach.* Asian Journal of Biomedical and Pharmaceutical Sciences, 2013. 3(18): p. 10-16.

[78] Di Martino, P., R. Di Cristofaro, C. Barthelemy, et al., *Improved compression properties of propyphenazone spherical crystals.* International journal of pharmaceutics, 2000. 197(1): p. 95-106.

[79] Mattox, D.M., *Handbook of physical vapor deposition (PVD) processing*2010: William Andrew.

[80] Mersmann, A., *Crystallization technology handbook.* Drying Technology, 1995. 13(4): p. 1037-1038.

[81] Burt, H.M. and A. Mitchell, *Effect of habit modification on dissolution rate.* International journal of pharmaceutics, 1980. 5(3): p. 239-251.

[82] Argon, A.S., *Strengthening mechanisms in crystal plasticity*2008: Oxford University Press Oxford.

[83] Thompson, C., *Investigating the fundamentals of drug crystal growth using Atomic Force Microscopy*, 2003, University of Nottingham.

[84] Brown, G., *Liquid crystals and biological structures*2012: Elsevier.

[85] Lin, W.-Q., J.-H. Jiang, H.-F. Yang, et al., *Characterization of chloramphenicol palmitate drug polymorphs by Raman mapping with multivariate image segmentation using a spatial directed agglomeration clustering method.* Analytical chemistry, 2006. 78(17): p. 6003-6011.

[86] Wells, A.F., *Structural inorganic chemistry*2012: Oxford University Press.

[87] Byrn, S., R. Pfeiffer, G. Stephenson, D. Grant, and W. Gleason, *Solid-state pharmaceutical chemistry.* Chemistry of materials, 1994. 6(8): p. 1148-1158.

[88] Borg, R.J. and G.J. Dienes, *An introduction to solid state diffusion*2012: Elsevier.

[89] Dunitz, J.D. and J. Bernstein, *Disappearing polymorphs.* Accounts of chemical research, 1995. 28(4): p. 193-200.

[90] Mirza, S., I. Miroshnyk, J. Heinämäki, et al., *Crystal morphology engineering of pharmaceutical solids: tabletting performance enhancement.* AAPS PharmSciTech, 2009. 10(1): p. 113-119.

[91] Raval, M.K., K.R. Sorathiya, N.P. Chauhan, et al., *Influence of polymers/excipients on development of agglomerated crystals of secnidazole by crystallo-co-agglomeration technique to improve processability.* Drug development and industrial pharmacy, 2013. 39(3): p. 437-446.

[92] Jolly, C.M., P. Lekshmi, I. Constantine, et al., *Crystallo-Co-Agglomeration: An Innovative Technique for Size Enlargement and Improved Flow Properties of Powders.* RESEARCH & REVIEWS: JOURNAL OF MATERIAL SCIENCES, 2013. 1(2): p. 1-14.

[93] Khougaz, K. and S.D. Clas, *Crystallization inhibition in solid dispersions of MK-0591 and poly(vinylpyrrolidone) polymers.* J Pharm Sci, 2000. 89(10): p. 1325-34.

[94] Chiang, P.-C., Y. Ran, K.-J. Chou, et al., *Evaluation of drug load and polymer by using a 96-well plate vacuum dry system for amorphous solid dispersion drug delivery.* AAPS PharmSciTech, 2012. 13(2): p. 713-722.

[95] Byrappa, K., R.I. Walton, A. Huang, et al., *Progress in crystal growth and characterization of materials.* Progress in Crystal Growth and Characterization of Materials, 2012. 58: p. 164-165.

[96] Montoya, E., A. Lorenzo, and L. Bausa, *Optical characterization of crystals.* Journal of Physics: Condensed Matter, 1999. 11(1): p. 311.

[97] Yadav, A., N. Bhagat, P. Mastud, and R. Khutale, *An Overview of Optimization of Spherical Crystallisation Process.* International Journal of Pharmaceutical Sciences and Nanotechnology, 2013. 8(4).

[98] Sandler, N. and D. Wilson, *Prediction of granule packing and flow behavior based on particle size and shape analysis.* Journal of pharmaceutical sciences, 2010. 99(2): p. 958-968.

[99] Rasenack, N. and B.W. Müller, *Ibuprofen crystals with optimized properties.* International journal of pharmaceutics, 2002. 245(1): p. 9-24.

[100] Katta, J. and Å.C. Rasmuson, *Spherical crystallization of benzoic acid.* International journal of pharmaceutics, 2008. 348(1): p. 61-69.

[101] Niefind, K., B. Riebel, J. Muller, W. Hummel, and D. Schomburg, *Crystallization and preliminary characterization of crystals of R-alcohol dehydrogenase from Lactobacillus brevis.* Acta Crystallographica Section D: Biological Crystallography, 2000. 56(12): p. 1696-1698.

[102] Hammouda, Y., L. El-Khordagui, I. Darwish, and A. El-Kamel, Manipulation of powder characteristics by interactions at the solid–liquid interface: 1-sulphadiazine. European journal of pharmaceutical sciences, 1999. 8(4): p. 283-290.